The Role of Vitamins in Combating Infectious Viral Diseases

This book highlights the role of vitamins in preventing or reducing the pathogenesis or treatment of infectious viral diseases based on current ongoing research and past work. Using clinical evidence and trials that suggest the potential benefits of vitamin supplementation as prophylactic and therapeutic in infectious viral diseases, each individual vitamin is described in this context in separate chapters. It will be a valuable reference aid to researchers, clinicians, and medical bodies to develop improved therapeutic regimens.

Key Features:

- Acts as a one-stop resource on the relevance of vitamins in infectious viral diseases.
- Provides a clinical focus on disease prevention and therapy using vitamins for clinicians and researchers.
- Discusses the molecular mechanisms of vitamins in COVID-19 and other viral diseases.

The Role of Vitamins in Combating Infectious Viral Diseases

Edited by
Aditya Arya and Rakesh Kaushik

CRC Press
Taylor & Francis Group
Boca Raton London New York

CRC Press is an imprint of the
Taylor & Francis Group, an **informa** business

Designed cover image: shutterstock.com/image-vector/medical-vitamin-mineral-background-multivitamin-complex-1091174246

First edition published 2025
by CRC Press
2385 NW Executive Center Drive, Suite 320, Boca Raton FL 33431

and by CRC Press
4 Park Square, Milton Park, Abingdon, Oxon, OX14 4RN

CRC Press is an imprint of Taylor & Francis Group, LLC

ISBN: 9781032564715 (hbk)
ISBN: 9781032563275 (pbk)
ISBN: 9781003435686 (ebk)

DOI: 10.1201/9781003435686

Typeset in Minion
by codeMantra

My Loving Grandfather, Sh. Chhedalal Arya, whom we lost while the book was being written. He survived the COVID-19 pandemic at the age of 80+ owing to his healthy and active lifestyle. Perhaps the diet and activity levels play an important role in longevity.

Contents

Preface

In recent years, the world has faced the formidable challenge of infectious viral diseases, leaving an indelible mark on global health, economies, and societies. From the emergence of groundbreaking coronaviruses like COVID-19 to the resurgence of formidable foes such as MERS, SARS, and Ebola, the urgency to address the threat posed by infectious diseases has never been more apparent. The current decade has witnessed the sad reality of approximately 56.2 million individuals succumbing to various diseases worldwide, with nearly one-third (26.1%) of these fatalities attributed to infectious diseases predominantly concentrated in developing regions. Among the staggering total of 14.7 million deaths recorded due to infectious causes, an overwhelming majority (14.2 million) occurred in these vulnerable areas. As multidimensional research accelerates its pace in diagnosis, vaccine development, and therapeutic drug discovery, the call for prophylaxis remains resounding. Against the backdrop of uncertainty in drug development and the pressing scale of time, one avenue of research stands out for its promising potential in combating these viral adversaries: the role of vitamins. Scientists delving into the intricate mechanisms of viral infections and immune responses have unveiled the therapeutic and prophylactic aspects of vitamins as a beacon of hope in our fight against infectious diseases. *The Role of Vitamins in Combating Infectious Viral Diseases* encapsulates the time-tested work of several researchers in a captivating and concise text. This critical confluence of nutrition and infectious disease zeros in on the role of vitamins in countering viruses that have wreaked havoc on human populations worldwide. From the unprecedented spread of COVID-19 to the persistent threats posed by other viral pathogens, this book explores the therapeutic and prophylactic potential of vitamins, aiming to illuminate their influence on immune responses, biochemical perturbations, and disease outcomes. Readers will find comprehensive discussions on the therapeutic benefits of various vitamins, supported by research evidence associated with viral infectious diseases such as MERS, SARS, COVID-19, and Ebola. This book is not merely a repository of knowledge but an invitation to embark on a scholarly exploration. It aims to stimulate intellectual curiosity, foster scholarly discourse, and encourage proactive engagement in enhancing personal health and resilience amid viral adversities. Together, readers are urged to uncover the latent capabilities of vitamins in combating infectious viral diseases. We firmly believe that this book serves as a timely and indispensable resource, providing a deeper understanding of the pivotal role that vitamins play in our collective defense against infectious threats. While the discussions within encompass opinions and evidence derived from clinical studies and trials, we must emphasize that our intention is not for readers to directly apply the text as clinical guidance. We extend our heartfelt gratitude to CRC Press for their kind consideration of the proposal and for bringing this book to our readers in its present form.

Warm regards,
Aditya Arya and Rakesh Kaushik

Acknowledgments

Editors are indebted to the Indian Council of Medical Research (ICMR) for project support. Dr Rakesh Kaushik would like to acknowledge the Indian Council of Agriculture Research for generous funding and research facilities. Dr Aditya Arya would like to acknowledge the support provided by his wife Sneha Singh in compiling the literature and support in preparation of some of the illustrations of the chapters while the book was being written.

Contributors

Nasreen Akhtar
Scientist-II, Vector Biology Department
National Institute of Malaria Research
Dwarka Delhi, India

Aditya Arya
Covid Diagnostic Laboratory
National Institute of Malaria Research
Delhi, India

Priya Chouhan
Department of Biochemistry, Division of Life Sciences
School of Basic & Applied Sciences
Galgotias University
Greater Noida, Uttar Pradesh, India

Ahana Dasgupta
Eicher Shroff Centre for Stem Cells Research (ES-CSCR)
Dr Shroff's Charity Eye Hospital
New Delhi, India

Vivek Dhar Dwivedi
Research Division
Quanta Calculus
Greater Noida, Uttar Pradesh, India

Meghna Garg
Department of Biotechnology
GLA University
Mathura, Uttar Pradesh, India

Anjana Goel
Department of Biotechnology
Institute of Applied Sciences & Humanities
GLA University
Mathura, Uttar Pradesh, India

Bhavya Jain
Cell and Developmental Biology Lab
School of Life Sciences
Jawaharlal Nehru University
Delhi, India

Sakthivel Jeyaraj
Centre for Drug Discovery and Development
Sathyabama Institute of Science and Technology
Chennai, Tamil Nadu, India

Rakesh Kaushik
Molecular Genetic Division
ICAR-Central Institute for Research on Goats (CIRG)
Makhdoom, Farah, Mathura, Uttar Pradesh, India

Lokesh Kori
Parasite Host Biology Division
National Institute of Malaria Research
Dwarka, Delhi, India

Shilpa Mahajan
Molecular Genetics Division
ICAR-Central Institute for Research on Goats
Makhdoom, Farah, Mathura, Uttar Pradesh, India

Rajan Malhotra
Department of Nutrition and Dietetics
Faculty of Allied Health Sciences
Manav Rachna International Institute of Research Studies
Faridabad, Haryana, India

Shashi Dhar Mehta
Department of Pulmonary Medicine
Vallabhbhai Patel Chest Institute Delhi
Delhi, India

Krupakar Parthasarathy
Centre for Drug Discovery and Development
Sathyabama Institute of Science and Technology
Chennai, Tamil Nadu, India

Ratna Rabha
Department of Chemistry
Banaras Hindu University
Varanasi, Uttar Pradesh, India

Shilpa Raina
Department of Applied Sciences
Shri Venkateshwara University
Gajraula, Uttar Pradesh, India

Sam Ebenezer Rajadas
Centre for Drug Discovery and Development
Sathyabama Institute of Science and Technology
Chennai, Tamil Nadu, India

Gurmeen Rakhra
Department of Biochemistry
School of Bioengineering and Biosciences
Lovely Professional University
Phagwara, Punjab, India

Gurseen Rakhra
Department of Nutrition and Dietetics
School of Allied Health Sciences
Manav Rachna International Institute of Research Studies
Faridabad, Haryana, India

T. Ramya
Department of Biological Sciences
Pilani Hyderabad Campus
Hyderabad, India

Sudhanarayani S. Rao
Centre for Drug Discovery and Development
Sathyabama Institute of Science and Technology
Chennai, Tamil Nadu, India

Jyoti Sangwan
Eicher Shroff Centre for Stem Cells Research
Dr Shroff's Charity Eye Hospital
New Delhi, India

Ayushi Sharma
Department of Biotechnology
Institute of Applied Sciences & Humanities
GLA University
Mathura, Uttar Pradesh, India

Vipin Kumar Sharma
Department of Applied Sciences
Shri Venkateshwara University
Gajraula, Uttar Pradesh, India

Nida Siddiqui
Covid-Diagnostic Laboratory
National Institute of Malaria Research (NIMR-ICMR)
Delhi, India

Himmat Singh
Vector Biology Department
National Institute of Malaria Research
Dwarka, Delhi, India

Sneha Singh
Department of Biotechnology
Jamia Hamdard University
New Delhi, India

Vignesh Sounderrajan
ICMR RA III, Centre for Drug Discovery and Development
Sathyabama Institute of Science and Technology
Chennai, Tamil Nadu, India

T. Thangam
Centre for Drug Discovery and Development
Sathyabama Institute of Science and Technology
Chennai, Tamil Nadu, India

Tarun Tyagi
Department of Internal Medicine
Yale School of Medicine
New Haven, Connecticut

Editors' Biographies

Aditya Arya, PhD is a scientist at the National Institute of Malaria Research, a flagship lab of ICMR, where he has main responsibilities of molecular diagnostics. He earned his PhD from the Peptide and Proteomics Division of the Defence Institute of Physiology and Allied Science, New Delhi, with proteomics and nanomedicine as thrust areas. Dr. Arya previously completed his Master's in Biochemistry with distinction merit from Madurai Kamaraj University. He has made a vast contribution to high-altitude biology, the discovery of proteomics signatures, and redox regulation in hypoxic-stress proven and recognized by 40+ research papers, 10+ book chapters, and 10+ books. He holds proven expertise in proteomics, microscopy, and cell-based biochemical assays, including flow cytometry, protein–protein interactions, and systems biology on which he acquired training from prestigious institutions such as European Molecular Biology Laboratory, Heidelberg; Wellcome Sanger Institute, Hinxton; Charite University, Berlin; and many more. He has also been awarded several grants and awards from EMBL, CSIR, and ICMR to present his work in more than ten countries. His current area of research focuses on developing a panel signature of integrated omics for various environmental stresses and the use of bioinformatics and proteomics approaches to obtain consensus anticancer descriptors from dietary antioxidants.

Rakesh Kaushik, PhD is a research associate at the flagship laboratory of ICAR, the Central Institute for Research on Goats. Before this, he held the position of Scientist-B at ICMR-National Institute of Malaria Research in New Delhi, focusing primarily on molecular diagnostics. Dr Kaushik earned his Doctorate in Biotechnology from GLA University, where he conducted experimental research at the Molecular Genetics Division of ICAR-Central Institute for Research on Goats (ICAR-CIRG) in Mathura. Delving extensively into the consequences of heat stress on biochemical, DNA, and RNA aspects, he has authored over 25 research papers, in addition to more than 35 abstracts, six book chapters, and a practical handbook on the subject. Dr Kaushik possesses extensive expertise in molecular techniques, including RT-PCR, 2-D electrophoresis, SDS-PAGE, and molecular marker analysis. During the challenging times of the COVID-19 pandemic, Dr Kaushik, in collaboration with ICMR, exhibited leadership by overseeing molecular diagnostic procedures for more than 50,000 samples.

Introduction

INFECTIOUS DISEASES AND VITAMINS

As readers dig further into this book, it will be crucial to review this foundational introduction on infectious diseases and vitamins, especially for those readers outside the realm of infectious biology and biochemistry. An **infectious disease** refers to "an ailment caused by a pathogen or its harmful substance, transmitted from an infected person, animal, or contaminated object to a vulnerable host." Infectious diseases, stemming from pathogenic microorganisms like bacteria, viruses, fungi, or parasites, pose significant health risks as they can spread directly or indirectly between individuals or from animals to humans. On these grounds, infectious diseases may be classified as bacterial-, viral-, fungal-, or parasitic-borne. Bacterial infections, induced by pathogens like *Escherichia coli* and *Streptococcus*, manifest in conditions such as pneumonia and tuberculosis. Fungal infections, instigated by organisms like *Candida* and *Aspergillus*, often target weakened immune systems or damp environments, leading to maladies such as athlete's foot and yeast infections. Parasitic infections, including malaria and giardiasis, result from parasites transmitted through vectors or contaminated food and water. Viral infections, caused by agents like influenza and COVID-19, range from mild colds to severe respiratory ailments by hijacking host cells for replication. Transmission modes encompass direct and indirect contact, vector-borne routes, and food and water-borne avenues.

Viral infections rank among the most prevalent ailments affecting humans. It is estimated that children endure between two to seven respiratory infections annually, while adults typically experience one to three such occurrences. Viruses are responsible for familiar infectious diseases from the common cold, influenza, and warts, to severe conditions such as HIV/AIDS, Ebola, and COVID-19. While millions of virus variants may exist, researchers have identified only around 5,000 types to date. Encapsulated within a protective coat of protein and lipid molecules, viruses contain a small genetic code. Upon invading a host cell, viruses release their genetic material, compelling the cell to replicate the virus, thereby leading to its multiplication. Subsequently, the infected cell may either release virus replicates to infect new cells or undergo functional alterations, potentially resulting in conditions like cancer induced by viruses such as human papillomavirus (HPV) and Epstein–Barr virus (EBV). Viruses, unlike many bacteria, need living cells to reproduce. This means they rely on a series of infections, called a chain of transmission, to survive in nature. Disease occurrence isn't always necessary or helpful for viruses. While visible infections may produce more infectious viruses, unnoticed infections are more common and unlikely to limit the movement of infected individuals, so they spread viruses widely. Epidemiologists identify three virus survival patterns in mammals based on reservoirs: short-term infections without a reservoir, long-term infections with a human reservoir, and infections involving an animal reservoir. The Baltimore classification, among various comprehensive viral classification schemes, categorizes viruses into multiple groups, including dsDNA (double-stranded DNA), ssDNA (single-stranded DNA), dsRNA (double-stranded RNA), positive-sense ssRNA (single-stranded RNA), negative-sense ssRNA, ssRNA-RT (single-stranded RNA with reverse transcriptase activity), and dsDNA-RT (double-stranded DNA with reverse transcriptase activity). Generally, DNA viruses replicate inside the cell nucleus, while RNA viruses replicate in the cytoplasm. However, there are exceptions to this rule in which poxviruses replicate in the cytoplasm and certain RNA viruses like orthomyxoviruses and hepatitis D virus replicate in the nucleus.

Viral infectious diseases constitute a major global health concern due to their widespread prevalence, significant morbidity and mortality rates, economic burden, and impact on global health security. These diseases affect millions of people annually, ranging from common illnesses like influenza and the common cold to more severe conditions such as HIV/AIDS, Ebola, Zika, and COVID-19. COVID-19 alone, at the time of the compilation of this book, has caused 7,003,621 deaths globally, while 703,876,819 people were infected globally. As we, the present generation have already witnessed the socio-economic impact of COVID-19, it can be assumed that most of the infectious diseases that reach a scale of pandemic or remain prevalent for longer periods, impose substantial economic costs through medical expenses, lost productivity, and disruptions to trade and commerce. Managing viral infections requires considerable healthcare resources and international collaboration to detect outbreaks early, contain their spread, and develop effective interventions. Addressing these challenges necessitates a comprehensive approach encompassing prevention, surveillance, treatment, and research efforts at local, national, and global levels. Preventive measures such as vaccination, hygiene practices, sanitation, vector control, and antimicrobial treatments, along with public health interventions such as surveillance and education, are paramount to curbing infectious disease spread locally and globally.

Treatment for most viral infections primarily focuses on alleviating symptoms as the immune system combats the virus. Some viral infections can be treated with antiviral medications, with notable advancements in antiviral therapeutics currently underway. Vaccination serves as a preventive measure against numerous viral diseases, which may lay dormant for a period before reactivating and causing illness again, even after the individual appears to have fully recovered. Vitamins, which are "vital" for health and known to be indispensable components of nutrition, are also interesting prophylactic agents as well as molecules of well-being. Vitamins play a crucial role in supporting the immune system's response to viral infections.

Vitamins are essential organic compounds required by the body in small amounts to maintain various physiological functions. They are classified into two main categories: **lipid-soluble vitamins** and **water-soluble vitamins**. Lipid-soluble vitamins include vitamins A, D, E, and K, which dissolve in fats and are stored in the body's fatty tissues and liver. These vitamins are absorbed along with dietary fats. In contrast, water-soluble vitamins, including the B-complex vitamins (B1, B2, B3, B5, B6, B7, B9, and B12) and vitamin C, dissolve in water and are not stored in the body to the same extent as fat-soluble vitamins. The B-complex vitamins are widely distributed in various foods. For instance, vitamin B1 (thiamine) can be found in whole grains, nuts, and legumes, while vitamin B2 (riboflavin) is abundant in dairy products, leafy greens, and eggs. Nuts, seeds, and fish are excellent sources of vitamin B3 (niacin), while vitamin B6 (pyridoxine) can be obtained from bananas, potatoes, and poultry. Vitamin B12 (cobalamin) is mainly found in animal products such as meat, fish, dairy, and eggs. These water-soluble vitamins are vital for numerous physiological functions, including energy metabolism, nerve function, and cell growth and repair. Vitamin C, found abundantly in citrus fruits like oranges, lemons, and grapefruits, as well as in vegetables like bell peppers, broccoli, and strawberries, plays a crucial role in immune function, collagen production, and wound healing (Figure i.1).

Fat-soluble vitamins are essential nutrients that dissolve in fat and are stored in the body's fatty tissues. There are four main fat-soluble vitamins: vitamins A, D, E, and K, each with unique functions and sources. Vitamin A is crucial for maintaining healthy vision, immune function, and skin health. It is abundant in foods like liver, fish liver oils, and dairy products, as well as colorful fruits and vegetables like carrots, sweet potatoes, and spinach. Vitamin D plays a vital role in regulating calcium and phosphorus absorption, promoting bone health, and supporting immune function. While the body can synthesize vitamin D when exposed to sunlight, it is also found in fatty fish (such as salmon and mackerel), egg yolks, and fortified foods like milk and cereal. Vitamin E acts as an antioxidant, protecting cells from damage caused by free radicals, and plays a role in immune function and skin health. It is present in nuts, seeds, vegetable oils (such as wheat germ, sunflower, and safflower oil), and green leafy vegetables. Vitamin K is essential for blood clotting, bone metabolism, and cardiovascular health. It is found in high amounts in green leafy vegetables like kale, spinach, and broccoli, as well as in some oils like soybean and canola oil.

Recommended dietary allowance (RDA) and adequate intake (AI) are both important guidelines used to determine the optimal intake of vitamins and other essential nutrients (such as minerals) for individuals. The RDA represents the average daily dietary intake level that is sufficient to meet the nutrient requirements of nearly all healthy individuals within a specific age and gender group, typically covering 97%–98% of the population. RDAs are established based on extensive scientific evidence, including studies on nutrient

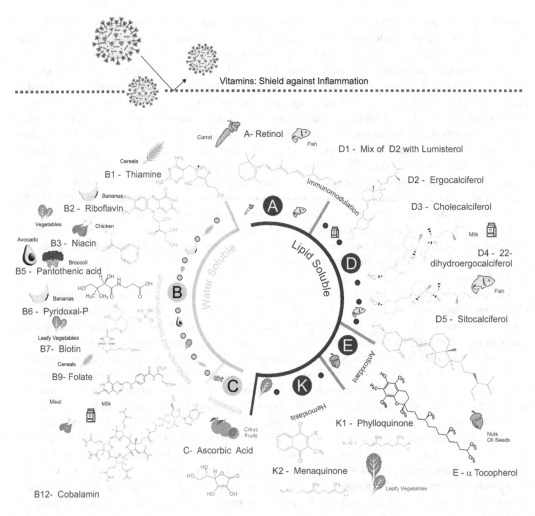

Figure i.1 Water-soluble and lipid-soluble vitamins provide a shield for inflammation, Vitamers and their natural sources.

requirements, absorption, metabolism, and bioavailability. They are expressed in specific units such as micrograms, milligrams, or international units and may vary by age, gender, life stage, and population groups. In contrast, AI is utilized when there is insufficient scientific evidence to establish an RDA. AI values represent an estimated daily intake level of a nutrient assumed to be adequate based on observed or experimentally determined intake levels in healthy populations. AI values are set at levels believed to cover the needs of most individuals within a specific population group, but they may also vary depending on age, gender, life stage, and population characteristics. Both RDAs and AIs are essential tools in dietary guidance, helping individuals achieve optimal nutrient intake for overall health and well-being. However, variations in RDAs and AIs may occur from region to region due to differences in dietary guidelines, nutritional recommendations, and population characteristics, reflecting the specific needs and considerations of diverse populations worldwide.

Recommended doses of vitamins are typically expressed in units (IU), micrograms (µg), or milligrams (mg), depending on the vitamin and the country's guidelines. International units (IU) are commonly used for lipid-soluble vitamins like vitamins A, D, E, and K. To interconvert vitamin doses between units (IU) and other measurements like micrograms (µg) or milligrams (mg), it's essential to understand the conversion factors for each vitamin. These conversion factors vary depending on the specific vitamin and its form (e.g., retinol for vitamin A, cholecalciferol for vitamin D). Here are the conversion factors for some commonly measured vitamins:

- Vitamin A: 1 IU of vitamin A is equal to 0.3 μg of retinol or 0.6 μg of beta-carotene. Conversely, 1 μg of retinol is equivalent to 3.33 IU of vitamin A.
- Vitamin D: 1 IU of vitamin D is equal to 0.025 μg (25 nanograms) of cholecalciferol (vitamin D3) or ergocalciferol (vitamin D2). Conversely, 1 μg of cholecalciferol is equivalent to 40 IU of vitamin D.
- Vitamin E: 1 IU of vitamin E is equal to 0.67 mg of d-alpha-tocopherol. Conversely, 1 mg of d-alpha-tocopherol is equivalent to 1.49 IU of vitamin E.
- Vitamin K: For vitamin K1 (phylloquinone), 1 IU is approximately equal to 0.03 μg. For vitamin K2 (menaquinone), conversion factors may vary depending on the specific form.

Conversion factors for vitamin B-complex vitamins (e.g., B1, B2, B3, B5, B6, B7, B9, B12) to IU are not typically used since these vitamins are water-soluble and do not have a standardized IU measurement. Instead, they are usually expressed in micrograms (μg) or milligrams (mg) based on their specific form (e.g., thiamine for B1, riboflavin for B2, niacin for B3, etc.). The RDAs or AI values are provided in these units. However, for folate or vitamin B9, sometimes conversion may be needed. In order to do so, divide value in micrograms by 0.6 to see the value as mcg DFE (Dietary Folate Equivalents) applies to folic acid and 5-MTHF. Similar to vitamin B, conversion factors for vitamin C (ascorbic acid) to IU are not commonly used because it is a water-soluble vitamin. Vitamin C is typically expressed in milligrams (mg) or grams (g) in dietary guidelines and supplements. However, if needed, the conversion factor can be calculated based on the molecular weight of ascorbic acid and the definition of IU, which varies depending on the country and regulatory body. Table i.1 provides a summary of recommended daily doses and common sources of vitamins.

Table i.1 Common vitamins, various forms (vitamers), daily recommended doses, and common food sources

Vitamin	Recommended daily doses*	Animal-based sources	Plant-based sources
A	900 μg (M) 700 μg (F)	Fish, liver, dairy products	Orange and ripe yellow fruits, leafy vegetables, carrots, pumpkin, squash, spinach
Chemical name and forms (vitamers): all-trans-retinol (retinal, retinoic acid, retinoids), provitamin A carotenoids (alpha-carotene, beta-carotene, gamma-carotene), xanthophyll beta-cryptoxanthin			
B1	1.2 mg (M) 1.1 mg (F)	Pork, liver, eggs	Wholemeal grains, brown rice, vegetables, potatoes
Chemical name and forms (vitamers): thiamine, thiamine monophosphate, thiamine pyrophosphate			
B2	1.3 mg (M) 1.1 mg (F)		Dairy products, bananas, green beans, asparagus
Chemical name and forms (vitamers): riboflavin, flavin mononucleotide (FMN), flavin adenine dinucleotide (FAD)			
B3	16 mg (M) 14 mg (F)	Meat, fish, eggs	Green Leafy vegetables, mushrooms, tree nuts
Chemical name and forms (vitamers): nicotinic acid, niacinamide, nicotinamide riboside			
B5	5 mg (M) 5 mg (F)	Meat	Broccoli, avocados
Chemical name and forms (vitamers): pantothenic acid, panthenol, pantethine			
B6	1.3–1.7 mg (M) 1.2–1.5 mg (F)	Meat	Vegetables, tree nuts, bananas
Chemical name and forms (vitamers): pyridoxine, pyridoxamine, pyridoxal			
B7	30 μg (M) 30 μg (F)	Raw egg yolk, liver,	Peanuts, leafy green vegetables
Chemical name: biotin			
B9	400 μg (M) 400 μg (F)	Liver	Leafy vegetables, pasta, bread, cereal,
Chemical name: folates, folic acid			

(Continued)

Table i.1 (*Continued*) Common vitamins, various forms (vitamers), daily recommended doses, and common food sources

Vitamin	Recommended daily doses*	Animal-based sources	Plant-based sources
B12	2.4 µg (M) 2.4 µg (F)	Meat, poultry, fish, eggs	Milk
Chemical name and forms (vitamers): cyanocobalamin, hydroxocobalamin, methylcobalamin, adenosyl cobalamin			
C	90 mg (M) 75 mg (M)	Liver	Citrus fruits (amla, orange) and vegetables
Chemical name: Ascorbic acid			
D1	15 µg		
Chemical name and forms (vitamers): mixture of molecular compounds of ergocalciferol with lumisterol, 1:1			
D2	–		Sunlight-exposed mushrooms and yeast
Chemical name: ergocalciferol			
D3	–	Fatty fish (mackerel, salmon, sardines), fish liver oils, eggs from hens fed vitamin D	
Chemical name: cholecalciferol			
D4	–	–	–
Chemical name: 22-dihydroergocalciferol			
D5	–	–	–
Chemical name: sitocalciferol			
E	15 mg	–	Many fruits and vegetables, nuts and seeds, seed oils
Chemical name and forms (vitamers): tocopherols, tocotrienols			
K1	110 µg (M) 120 µg (F)	–	Leafy green vegetables such as spinach
Chemical name: phylloquinone			
K2	110 µg (M) 120 µg (F)	Poultry and eggs, natto	–
Chemical name: menaquinone			

* In some cases, the recommended doses may be adequate dose of AI.

It's important to note that these conversion factors are approximate and may vary slightly depending on the source and assay used for measurement. Additionally, the recommended daily allowance (RDA) or recommended dietary allowance (RDA) for vitamins may be expressed in different units based on regional guidelines and recommendations. When determining vitamin doses and interconverting between units, it's essential to refer to reliable sources such as dietary guidelines, scientific literature, or healthcare professionals to ensure accurate dosing and adequate nutrient intake for optimal health.

We now hope that its good time to dive deeper into the concept of vitamins and their roles in combating infectious viral diseases.

Infectious viral diseases
History and pandemics

SNEHA SINGH AND TARUN TYAGI

INTRODUCTION

The global burden of diseases encompasses the overall impact of health conditions on populations worldwide, including mortality, morbidity, and disability. Infectious diseases play a significant role in this burden, particularly in low- and middle-income countries. These diseases, caused by pathogens such as bacteria, viruses, parasites, and fungi, contribute to a substantial portion of the global disease burden (Antabe et al., 2019). They can lead to severe illness, disability, and death and often disproportionately affect vulnerable populations with limited access to healthcare and sanitation. Examples of infectious diseases with high global burdens include human immunodeficiency virus (HIV)/acquired immunodeficiency syndrome (AIDS), tuberculosis, malaria, respiratory infections (such as influenza and pneumonia), diarrheal diseases, and neglected tropical diseases (Michaud, 2009). The Global Burden of Disease (GBD) study, led by the Institute for Health Metrics and Evaluation (IHME) at the University of Washington, is a comprehensive epidemiological survey offering vital insights into global health challenges (GBD-2020). Partnering with *The Lancet* and the World Health Organization, the GBD study aimed to enhance its validity and policy relevance. It reveals alarming trends, highlighting a global crisis of chronic diseases and failure to address preventable risk factors, leaving populations vulnerable to health emergencies like COVID-19. Covering data from 1990 to the present, it analyzes mortality, morbidity, and risk factors across 204 countries and territories, encompassing 369 diseases and injuries. This study serves as a crucial resource for clinicians, researchers, and policymakers, aiding in tracking progress and informing health strategies. The GBD study, as reported in *The Lancet*, provides comprehensive estimates of various health metrics, including incidence, prevalence, mortality, years of life lost (YLLs), years lived with disability (YLDs), and disability-adjusted life years (DALYs). Notably, since 2010, there has been a notable acceleration in the decline of global age-standardized DALY rates, particularly among age groups younger than 50 years, with the most significant annualized rate of decline observed in the 0–9-year age group. Among children under 10 years old in 2019, six infectious diseases featured among the top ten causes of DALYs, including lower respiratory infections, diarrheal diseases, malaria, meningitis, whooping cough, and congenital syphilis. In adolescents aged 10–24 years, three injury causes ranked among the top causes of DALYs, including road injuries, self-harm, and interpersonal violence. Ischemic heart disease and stroke were identified as the leading causes of DALYs in age groups 50–74 years and 75 years and older. The study also highlights a significant shift toward a greater proportion of burden due to YLDs from noncommunicable diseases and injuries since 1990 (GBD-2020; Horton, 2020). While these statistics are dynamic and likely to change over the period of time, the understanding of disease occurrence and management is another crucial aspect, and a distinction between the patterns such as outbreaks, pandemic, epidemic, and endemic diseases must be understood well.

An **outbreak** refers to a sudden increase in disease cases beyond the expected norm in a specific community or region (Riley, 2019). For instance, in December 2019, reports emerged of an outbreak of a novel respiratory illness in Wuhan, China, marking the early stages of what would become a global epidemic. An **epidemic** occurs when the outbreak spreads rapidly across a larger population. In the case of COVID-19, caused by the SARS-CoV-2 virus, what began as an epidemic in Wuhan quickly escalated into a pandemic as cases surged worldwide (Wang et al., 2021). A **pandemic** is characterized by widespread transmission of a disease across multiple countries or continents, overwhelming healthcare systems and causing significant societal disruption. Efforts to control the pandemic, including widespread vaccination campaigns, aim to transition the virus from a pandemic to an endemic state (Purohit et al., 2022; Yin et al., 2022). **Endemic** diseases, like the seasonal flu, are constantly present in a community but are typically manageable and do not lead to widespread disruption. The ultimate goal is to reach a state where COVID-19 becomes an endemic disease, allowing for more manageable control measures and a return to normalcy. In yet another study, *The Lancet* has proposed the term syndemic for COVID-19, stating that: "Within certain populations, there's an interaction between two disease categories: infection with severe acute respiratory syndrome coronavirus 2 (SARS-CoV-2) and various non-communicable diseases (NCDs). These illnesses tend to occur more frequently among specific social groups, reflecting the deep-rooted patterns of inequality prevalent in our societies." The clustering of these diseases within communities experiencing social and economic disparity worsens the negative impacts of each individual disease (Horton, 2020). A **syndemic** is thus defined by two or more illness states interacting poorly with each other and negatively influencing the mutual course of each disease trajectory.

If we carefully observe the history of major pandemics in time, we would notice that infectious diseases often turned into pandemics and such evidence is known from antiquity. Several major pandemics throughout history have claimed millions of lives, leaving a profound impact on global populations. One of the deadliest pandemics in history is the Black Death or the bubonic plague pandemic (McEvedy, 1988). It ravaged Europe, Asia, and Africa in the 14th century, killing an estimated 75–200 million people. The HIV/AIDS pandemic, which began in the 1980s, has also claimed millions of lives worldwide (Sharp & Hahn, 2011). More recently, the COVID-19 pandemic caused by the novel coronavirus SARS-CoV-2 has led to significant loss of life globally. While we will discuss these pandemics in greater detail in later part of the chapter, here we may prefer to note that while most of the pandemics earlier than the 19th century were of bacterial origin, more recently viral pandemics have become rampant. Also considering the focus of this book, we would limit most of our discussion to post-1900 pandemics which were caused primarily by viruses. Perhaps, discovery of antibiotics in the early 20th century resulted in a major blow to bacterial infections. Viral pandemics on the other hand seem to show a much rapid rising trend, in contrast to bacterial pandemics perhaps, in lack of equally powerful antivirals. With nearly 11,273 viral species known (as per ICTV), among which 270 species (within 26 families) are known to infect humans, albeit it is claimed that many viruses still remain to be discovered, and therefore, the list could be unexpectedly large in reality (Woolhouse et al., 2012; Mushegian, 2020; Forni et al., 2022). The ten most dangerous viruses include the Marburg virus, notorious for its 90% fatality rate and hemorrhagic fever symptoms; Ebola, with its five strains, the deadliest being Zaire Ebola, causing hemorrhagic fever and organ failure; Hantavirus, associated with lung disease, fever, and kidney failure; Bird flu virus (H5N1), posing a 70% mortality rate primarily through poultry contact; Lassa virus, transmitted by rodents and causing fever and organ failure; Junin virus, linked to Argentine hemorrhagic fever and presenting with tissue inflammation and bleeding; Crimea-Congo fever virus, transmitted by ticks and manifesting pin-sized bleedings; Machupo virus, causing Bolivian hemorrhagic fever with high fever and bleeding; Kyasanur forest virus (KFD), found in Indian woodlands and leading to fever, headaches, and bleeding; and Dengue fever, transmitted by mosquitoes and affecting millions in tropical regions with symptoms like fever and severe headache. These viruses pose significant health threats worldwide, necessitating vigilant surveillance and preventive measures.

Throughout history, several pandemics caused by viruses have inflicted significant devastation on global populations. The Spanish flu pandemic, occurring from 1918 to 1919, remains one of the deadliest, caused by the H1N1 influenza virus, infecting approximately one-third of the world's population and resulting in millions of deaths. The HIV/AIDS pandemic, which began in the 1980s and continues to affect millions worldwide, has claimed countless lives due to HIV. Other notable pandemics include the Asian flu (1957–1958) and the Hong Kong flu (1968–1969), both caused by influenza viruses, resulting in widespread illness and mortality. More recently, the H1N1 influenza pandemic of 2009–2010, also known as the swine flu, and the ongoing COVID-19 pandemic caused by the novel coronavirus SARS-CoV-2 have further emphasized the critical importance of robust public health responses and global cooperation in combating infectious diseases.

WHAT CAUSES PANDEMICS?

Several variables about the variety of diseases and how they interact with people can cause pandemics. Geographically based, the term "pandemic" refers to a range of events and public health hazards, each with unique disease features, frequency, and severity (Madhav et al., 2017). Pathogens vary in many ways, such as how diseases spread and how quickly they do so, the severity of the morbidities they cause, and how well they can distinguish between different symptoms. Certain infections, including pandemic influenza viruses, have a strong chance of causing widespread and severe pandemics because of their effective human-to-human transmission, protracted periods of asymptomatic infection, and difficult differential diagnosis (Moghadami, 2017). Some infections, such as the Nipah virus and some influenza strains, pose a moderate worldwide danger because they can evolve and adapt to become more effective. Furthermore, many viruses, such as Ebola and Marburg, possess the capacity to initiate regional or interregional outbreaks, but their likelihood of triggering a worldwide pandemic is reduced owing to their slower transmission rate or increased likelihood of discovery and containment (Ayouni et al., 2021). The primary concern among all pandemic infections is influenza because of its frequent incidence and possible severity. The destructive impact of pandemics is exemplified by historical occurrences such as the 1918 influenza pandemic, which caused enormous worldwide mortality. Furthermore, during pandemics, differences in mortality rates between high- and low-income nations are impacted by several variables, including interconnectivity between population centers, comorbid conditions, malnutrition, and access to healthcare. These factors highlight the vital significance of developing preparedness and response plans that are specific to the features of each pandemic threat.

The epidemiological triangle model of infectious disease etiology, which emphasizes the interaction between three essential elements—**the agent** (pathogen), **the susceptible host**, and **the environment**—can be used to understand the emergence of a pandemic (Snieszko, 1974). The main source of infection and illness is the agent, often known as the pathogen. Its capacity to infect a host is determined by factors including pathogenicity, virulence, and infectivity. Those who are prone to infection are referred to as susceptible hosts. Numerous elements, such as immunological response, underlying medical problems, and heredity, affect host vulnerability. The physical, social, behavioral, cultural, political, and economic elements that contribute to the interaction between the agent and the host and ultimately result in infection and disease transmission are all included in the environment. Population density, urbanization, globalization, infrastructure for healthcare, and resource accessibility are a few examples of these environmental variables (van Seventer & Hochberg, 2017). After the host is exposed to the pathogen, the agent and host interact during a number of phases that may lead to infection, the development of the disease, and eventually either recovery or death. Both the pathogen's characteristics and the host's sensitivity affect how one stage develops into the next. The immunological response of the host is important in deciding how the infection turns out since it may either protect the host from the virus or have negative repercussions. Thus, appreciating the intricate interplay among the agent, host, and environment is crucial to understanding the underlying reasons for pandemic recurrence and to devising practical preventive, mitigating, and control measures.

Modes of transmission for a pandemic viral agent can be classified into direct and indirect transmission routes (van Seventer & Hochberg, 2017). **Direct transmission** involves the immediate transfer of the infective agent from a reservoir or host to a susceptible individual through various means such as physical contact, respiratory droplets, environmental exposures, bites from infected animals, or transmission from mother to fetus during pregnancy or childbirth. In contrast, **indirect transmission** occurs when the infective agent is transferred indirectly from a reservoir or host to a susceptible individual through intermediary sources. This includes biological transmission, where vectors like arthropods or intermediate hosts play a role in transmitting the agent; mechanical transmission through vectors or inanimate objects carrying the agent; and airborne transmission where respiratory droplets containing the agent remain suspended in the air. Understanding these modes of transmission is crucial for implementing effective preventive measures and controlling the spread of infectious diseases. Furthermore, among the direct transmission, human-to-human transmission can further be subcategorized into various types, such as respiratory or salivary transmission (e.g., influenza, measles, rhinovirus), fecal-oral transmission (e.g., enteroviruses, rotavirus), sexual contact (e.g., HIV, herpes simplex virus, papillomavirus), bloodborne transmission (e.g., HPV, HBV, HCV), vertical transmission from mother to fetus (e.g., HBV, HIV, rubella, CMV, HSV, enterovirus), and human–arthropod–human transmission (e.g., dengue, sandfly fever) (Richard et al., 2017).

Apart from human-to-human transmission, pandemics can also originate from animal-to-human transmission, known as **zoonosis**. Zoonotic origins may involve vertebrate-to-vertebrate transmission (e.g., most arboviruses) or vertebrate reservoir-based transmission (e.g., rabies). Understanding these modes of transmission is crucial for implementing effective preventive measures and controlling the spread of viral agents during a pandemic (Figure 1.1). According to the World Health Organization (WHO), zoonoses are illnesses

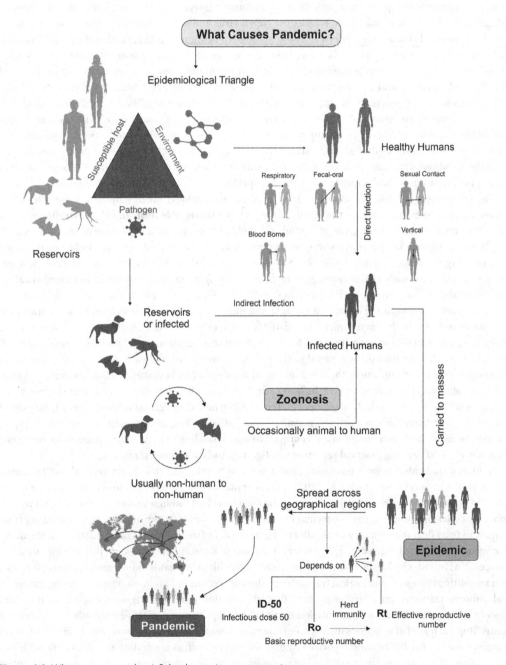

Figure 1.1 What causes pandemic? A schematic representation.

or infections that are naturally transmissible between vertebrate animals and humans or vice versa. The term is derived from the Greek terms "zoon," which means animal, and "nosos," which means illness (WHO). They make up around 61% of human infections, which makes them a serious public health problem. These illnesses directly endanger human health and frequently result in death. The 13 most prevalent zoonoses have an estimated 2.4 billion incidences of disease and 2.7 million fatalities yearly, harming animal health and decreasing livestock productivity (Grace et al., 2012). They particularly affect low- and middle-class livestock workers in these nations. Based on their cause, zoonoses are divided into several groups, such as bacterial (such as anthrax and salmonellosis), viral (such as rabies and Ebola), parasitic (such as toxoplasmosis and malaria), and fungal (Chomel, 2009).

Numerous variables impact pathogenesis. The way that agent determinants and host susceptibility interact determines how an infectious agent affects a potential host. Agent factors include pathogenicity, which describes the agent's capacity to cause disease once infection occurs, virulence, which indicates the probability of producing severe disease in individuals infected, and infectivity, which describes the possibility of the agent infecting a host upon exposure. The structural or biochemical characteristics of the infectious agent, such as the generation of toxins like the cholera toxin that causes watery diarrhea, are frequently the cause of virulence. Agent characteristics, including infectivity, are often measured using metrics like the **infectious dose 50 (ID50)**, representing the amount of agent needed to infect 50% of a specified host population (Yezli & Otter, 2011). Environmental factors, both physical and social behavioral, also play a role as extrinsic determinants of host vulnerability to exposure. Understanding these factors is crucial for elucidating the mechanisms of disease pathogenesis and developing strategies for prevention and treatment.

In addition to these factors, the sporadic nature and spread of a disease are assessed through the analysis of incidence and prevalence, which ultimately rely on the concepts of the reproduction number and herd immunity (McDermott, 2021). The **basic reproductive number (R_0)** serves as a gauge of the potential spread of an infectious disease within an immunologically naive population, representing the average number of secondary cases originating from a single infectious case in a fully susceptible population (Milligan & Barrett, 2015).

However, in reality, some individuals are usually immune (and therefore not susceptible) due to previous infection or immunization upon the introduction of most infectious diseases into a community. Thus, a more accurate measure of the potential for disease spread within a community is the effective reproductive number (R), which denotes the average number of new infections caused by a single infection (Scire et al., 2020). Generally, for an epidemic to develop within a population, the effective reproductive number must exceed 1, indicating that the number of cases continues to rise. Herd immunity, also known as community immunity, describes the resistance of a population to an infectious disease when a sufficient number of immune individuals exist to disrupt the chain of infection transmission (Ashby & Best, 2021). Consequently, susceptible individuals who lack immunity themselves are indirectly shielded from infection.

MAJOR GLOBAL PANDEMICS (PRE-20TH CENTURY)

The temporal canvas of human history unfolds with the documentation of 249 pandemics dating back to 1200 BC, persisting until the present epoch dominated by the COVID-19 virus. During antiquity, the obscured genesis of these pandemics stemmed from a limited medical cognizance, rendering the origins enigmatic. As we traverse the epochs, we dissect narratives depicting resilience, societal metamorphoses, and the progressive elucidation of medical paradigms that have shaped responses to these infectious phenomena. The **Antonine Plague**, a pivotal moment in the annals of the Roman Empire, erupted during the governance of Marcus Aurelius around 165 AD and persisted under the rule of his son, Commodus. This pandemic, also known as the **Plague of Galen**, played a transformative role in altering the pathocenosis of the Ancient World (Sabbatani & Fiorino, 2009). Its propagation was facilitated by two military campaigns in which Marcus

Aurelius actively participated—the Parthian War in Mesopotamia and conflicts against the Marcomanni in northeastern Italy, Noricum, and Pannonia. The scarcity of comprehensive clinical accounts adds a layer of complexity to our understanding, with Galen, a witness to the plague, providing fragmented insights primarily focused on therapeutic approaches rather than meticulous symptomatology. Although paleopathological confirmation of smallpox as the causative agent is absent, archaeological evidence, particularly terracotta artifacts from Italy, suggests a correlation. These artifacts reveal detailed depictions reminiscent of classic smallpox pustules, potentially affirming the artist's intent to represent the distinctive signs of the disease. The Antonine Plague introduced to Rome by returning armies from Western Asia, manifested in fever, skin sores, diarrhea, and sore throats. The absence of prior exposure and immunity among Roman citizens likely contributed to the catastrophic mortality rate, estimated at 25%. This devastating pandemic, believed to be linked to smallpox and measles, left an enduring mark on the trajectory of the Roman Empire, underscoring the profound impact of infectious diseases in antiquity. While smallpox has been eradicated, measles, characterized by its distinctive rash, persists today, claiming thousands of lives annually despite the availability of a preventive vaccine.

The catastrophic Justinian disease, also known as the **Justinianic Plague**, ravaged the Mediterranean region, Europe, and the Near East between 541 and 549 AD. Its epicenter was the Sasanian Empire and the Byzantine Empire, which had a significant influence on Constantinople. During that time, the pandemic peaked and killed about one-fifth of the population in the imperial capital, except for Justinian I, the Byzantine emperor, who acquired and survived the disease in 542 AD. The beginnings of the infection may be found in Roman Egypt in 541 AD, where it continued to spread over the Mediterranean until 544 AD and Northern Europe and the Arabian Peninsula until 549 AD. The Justinianic Plague, the first outbreak of the first plague pandemic, had significant political, social, and economic ramifications throughout Europe and the East. The name "Justinianic" is derived from Emperor Justinian's reign (r. 526–565 CE), with Constantinople as its imperial capital, where Justinian himself reportedly succumbed to the disease, although it did not prove fatal. As the beginning of the Middle Ages and the end of antiquity, it had a profound cultural and theological influence on Eastern Roman society. The parent of the Justinian plague strain found in the Tian Shan mountain ranges on the borders of Kyrgyzstan, Kazakhstan, and China is closely related to both ancient and present strains of *Yersinia pestis*. The bubonic plague is said to have originated in China, but reports of it date back to 610 CE. According to current mortality estimates, the Justinianic Plague killed tens of millions of people in the Mediterranean between around 541 and 750 CE.

Japan's first smallpox outbreak occurred in the 8th century, starting in 735 CE and leaving a lasting legacy in the country's history. This vicious epidemic, which killed off around one-third of the population, is known as a **virgin soil epidemic**, meaning that the people who experienced it had never been exposed to or developed an immunity to the smallpox virus (Suzuki, 2011). Following that, Japan saw 28 smallpox outbreaks, with a noticeable pattern of fewer time intervals between cases. The average time between smallpox outbreaks from 735 to 1000 CE was 24 years; however, from 1001 to 1206 CE, this period significantly decreased to 13 years. Smallpox in Japan changed from being an occasional breakout to an endemic condition as the Tokugawa period marked the beginning of the early modern era. Data from this era highlight the durability of smallpox and its effects on local populations, showing a pattern of significant outbreaks occurring in some villages around every 10 years. Of particular interest is the statistical finding that about 95% of smallpox-related fatalities involved people under the age of 10, underscoring the susceptibility of the younger demographic. Early in 1946, even as Japan was going through post-war reconstruction following World War II, a smallpox outbreak broke out (Tanaka et al., 2014).

The **Bubonic Plague**, famously known as the Black Death, emerged as a catastrophic pandemic during the mid-14th century, leaving an enduring mark on Europe, northern Africa, and the Near East (McEvedy, 1988; Duncan & Scott, 2005). In the year 1346 AD, a population of approximately 100 million inhabited this vast region. However, within 5 years, the relentless onslaught of the Black Death claimed the lives of 25 million individuals, signaling one of the most devastating pandemics in recorded history. The plague, caused by the bacterium *Y. pestis*, inflicted swift and horrific deaths upon its victims, manifesting as high fever accompanied by the painful emergence of suppurative buboes or swellings, which were characteristic of the disease. For much of the 20th century, it was widely believed that all plagues in Europe from 1347 to 1670 were manifestations of bubonic plague. The 14th-century outbreak alone claimed more than one-third of

Europe's population, translating to approximately 25 million lives lost. The causative agent, *Y. pestis*, has left an indelible mark on history, recurring in plague cycles from the Bronze Age to modern-day occurrences in regions like California and Mongolia. Despite advancements in medical science, plague remains endemic in certain areas such as Madagascar, Congo, and Peru, serving as a stark reminder of the enduring impact of this ancient and formidable infectious disease (Glatter & Finkelman, 2021).

The variola virus, which causes smallpox, provides evidence of the historical and worldwide effects of infectious illnesses on humankind. Smallpox was one of the worst illnesses when it was first identified in human populations; the variola major type has a 30% death rate (Thèves et al., 2014). Even while survivors developed protection against new illnesses, they nevertheless had to deal with lifetime effects. Edward Jenner's discovery of the first vaccination stages in the 18th century marked a turning point in the fight against this deadly illness. Following vaccination programs in the 19th and 20th centuries, the WHO made the historic announcement in 1980 that smallpox had been eradicated. This significant event happened around 2.5 years after Merca, Somalia, reported the last known naturally occurring instance (Strassburg, 1982). Its successful eradication marks a significant achievement in public health, making smallpox the first human infectious disease to be eliminated globally.

MAJOR GLOBAL PANDEMICS (20TH CENTURY TO PRESENT)

A terrible influenza pandemic known as the **Spanish flu** ravaged the world in 1918 and persisted until 1919. With an estimated third of the world's population infected and 50 million deaths, it continues to rank among the deadliest pandemics in history (Taubenberger & Morens, 2020). The H1N1 influenza A virus was the cause, and it was made more contagious and virulent by its segmented RNA genome and the glycoproteins hemagglutinin (H) and neuraminidase (N) on its surface (Kosik & Yewdell, 2019). Public health measures like isolation, quarantine, proper personal hygiene, the use of disinfectants, and restrictions on public meetings were put in place to stop the spread despite the lack of contemporary medical resources. However, the concurrent World War I hindered the worldwide reaction, with several nations downplaying the outbreak's seriousness to boost morale throughout the conflict. In order to prevent future pandemics, strong public health infrastructure and international collaboration are critical, as demonstrated by the way the Spanish flu was managed.

The **Hong Kong flu pandemic** occurred from 1968 to 1969, caused by the H3N2 influenza A virus (Jester et al., 2020). With 500,000 confirmed cases and an estimated 1–4 million fatalities globally, the virus possessed an RNA genome containing the glycoproteins neuraminidase (N) and hemagglutinin (H). Quarantine laws, vaccination drives, and the use of antiviral drugs were among the management techniques used to lessen the pandemic's effects.

The H2N2 influenza A virus was the cause of the 1957–1958 **Asian flu epidemic**. It caused one to 2 million fatalities worldwide (Menon, 1959; Schäfer et al., 1993). The virus combined genetic components from human and avian influenza viruses to create a novel structure. The creation of vaccines, isolation protocols, and antiviral drugs were among the management initiatives, underscoring the significance of international collaboration in combating infectious illnesses.

SARS, or **severe acute respiratory syndrome**, was a pandemic that started in Guangdong Province, China, and lasted from 2002 to 2003. The SARS coronavirus (SARS-CoV), a new virus in the Coronaviridae family, is the source of this viral outbreak (Cherry and Krogstad, 2024). The term "coronavirus" comes from the fact that the structure of the virus is a lipid envelope studded with spike proteins, which under a microscope gives it a crown-like look (Li et al., 2005; Pal et al., 2020). Around 8,000 cases were documented worldwide throughout the pandemic, resulting in 774 fatalities, or a 10% mortality rate. Scientists, governments, and health groups worked quickly to coordinate the global response. Tight public health protocols, including travel bans, isolation, and quarantine, were put in place to stop the virus's spread. The virus was quickly identified, thanks to research efforts, and cooperative projects made it possible to create efficient diagnostic tests. In the end, strict public health initiatives combined with advances in medical knowledge and treatment were able to contain the SARS outbreak.

The 2012 **Middle East respiratory syndrome (MERS)** epidemic mostly affected nations on the Arabian Peninsula. The causal agent was found to be the MERS coronavirus (MERS-CoV), which is related to the

SARS-CoV (severe acute respiratory syndrome coronavirus) family (Banik et al., 2015; de Wit et al., 2016). A lipid envelope packed with spike proteins that aid in host cell entrance is part of the MERS-CoV virus structure (Li et al., 2019). With 858 confirmed deaths and 2,494 confirmed cases overall, the epidemic had a roughly 34% mortality rate. Early public health interventions, such as case isolation, contact tracing, and increased surveillance, were critical to controlling the epidemic. Coordinating and communicating across international borders was essential to stopping the virus's spread. The MERS pandemic has demonstrated the value of international collaboration in responding to newly developing infectious diseases, while not being as prevalent as some other respiratory ailments.

Originating in West Africa, the **Ebola epidemic** was a terrible worldwide health disaster that mostly affected people between 2014 and 2016 (Del Rio & Guarner, 2015). The Ebola virus, which belongs to the Filoviridae family and is distinguished by its filamentous form, was the cause of the outbreak (Falasca et al., 2015). The virus mostly disseminated via direct contact with an infected person's body fluids, which can result in severe hemorrhagic fever, which has a high fatality rate. It was one of the worst Ebola outbreaks in history, with over 11,000 deaths and almost 28,000 recorded cases during the epidemic. Collaboration between several countries, research institutes, and health groups typified the global response. The main initiatives included sending out medical staff, setting up treatment facilities, enforcing stringent infection control protocols, and creating and dispensing experimental vaccinations. The pandemic was finally brought under control thanks to these coordinated efforts, underscoring the significance of international collaboration in controlling infectious disease epidemics.

The **2009 swine flu (H1N1) pandemic** was a worldwide health emergency brought on by the quick spread of a brand-new influenza A virus subtype. The virus was first discovered in pigs, but it changed to infect people and spread widely (Cohen, 2010). On June 11, 2009, the WHO proclaimed it to be a pandemic. Over 200,000 people had died and an estimated 1.4 billion people had been infected globally by the time the epidemic subsided in August 2010 (WHO-H1N1). The genetic composition of the H1N1 virus was distinct, including components from swine, avian, and human influenza strains. Hemagglutinin (H) and neuraminidase (N) proteins, which are essential for viral entrance and exit, were present on its surface (Garten et al., 2009). Managing the pandemic involved widespread vaccination campaigns, antiviral medications, public health measures like social distancing, and international collaboration to monitor and control the virus's spread.

Since its emergence in the late 20th century, the **HIV/AIDS pandemic** has presented a serious threat to world health. The first examples of homosexual males in the United States were documented in the early 1980s (Highleyman, 1999). The virus expanded around the world throughout the years, resulting in millions of cases and fatalities. Around 38 million individuals were estimated to be living with HIV/AIDS, and the epidemic has claimed over 36 million deaths, according to the latest estimates from the WHO and CDC (WHO-HIV). Retroviruses like HIV target the immune system, namely CD4 cells. Its structure is made up of an RNA genome, enzymes, and an outer envelope surrounded by glycoproteins (Turner & Summers, 1999). Significant strides have been achieved in controlling the epidemic through antiretroviral medication, education, and preventive initiatives, even though a cure has not yet been found. Global efforts, such as treatment accessibility and awareness campaigns, have been crucial in lowering the incidence of HIV/AIDS transmission and enhancing the quality of life for people who are infected (Figure 1.2).

The **COVID-19 pandemic**, caused by the novel coronavirus SARS-CoV-2, emerged in late 2019 in the city of Wuhan, Hubei Province, China (Cheng et al., 2020). The virus swiftly spread globally, leading to a profound impact on public health and economies with the first case in India by early 2020, the second most populous country (Andrews et al., 2020). By early 2022, there were millions of confirmed cases and significant fatalities worldwide. The WHO on March 11, 2020 declared the novel coronavirus (COVID-19) outbreak as a global pandemic, and India declared COVID-19 as a "notified disaster" on March 14, 2020 (WHO-COVID). The virus is a member of the Coronaviridae family, which is distinguished by its surface spikes that resemble crowns. Within a lipid membrane resides its genetic material, RNA. Widespread public health measures were triggered by the epidemic to stop transmission, such as mask laws, social separation, and lockdowns. The creation and dissemination of many vaccinations offered promise for limiting the disease's intensity and spread, and vaccination drives were essential in handling the crisis. Working together internationally, advancing science, and raising public awareness were essential in overcoming the obstacles the COVID-19 epidemic presented.

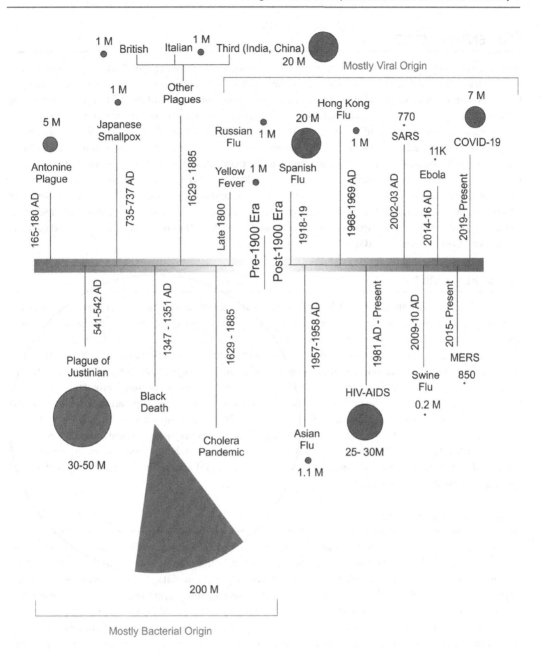

Figure 1.2 Major global pandemics timeline, pre-1900 era and post-1900 era (predominated by viral infectious diseases).

MANAGEMENT OF THE VIRAL PANDEMICS: LESSON LEARNT FROM THE HISTORY

Pandemic management involves a combination of preventive measures, public health interventions, medical treatments, and societal responses aimed at controlling the spread of infectious diseases and mitigating their impact on individuals and communities. The following are some key means of pandemic management.

PREVENTIVE MEASURES

Preventive measures in pandemic management encompass a multifaceted approach aimed at reducing the risk of virus transmission and protecting public health. This includes widespread vaccination campaigns to bolster immunity against the virus and limit its spread within communities. Additionally, promoting basic hygiene practices such as regular handwashing with soap and water, practicing respiratory etiquette by covering coughs and sneezes, and avoiding touching the face can significantly reduce the likelihood of virus transmission. Implementing physical distancing measures, such as maintaining a safe distance from others and reducing the size of gatherings, helps to minimize close contact and mitigate the spread of the virus. The requirement to wear a mask in public spaces further provides a barrier against respiratory droplets carrying the virus. Travel restrictions, including quarantine protocols and testing requirements, are instrumental in preventing the introduction and spread of the virus across regions. By adhering to these preventive measures collectively, individuals and communities can contribute to breaking the chain of transmission and controlling the spread of the pandemic (Güner et al., 2020).

BIOSAFETY MEASURES

Biosafety measures are critical safeguards implemented to prevent the accidental release or exposure to harmful biological agents in laboratory settings and other environments where infectious materials are handled. These measures encompass a range of protocols, practices, and infrastructure aimed at minimizing the risk of laboratory-acquired infections and ensuring the safety of laboratory personnel, the community, and the environment. Key components of biosafety include the classification of biological agents based on their potential risk, the implementation of appropriate containment measures, the use of personal protective equipment (PPE) such as gloves, goggles, and face masks, and the establishment of standard operating procedures (SOPs) for safe handling, storage, and disposal of biological materials. Biosafety also involves regular training and education of laboratory staff on risk mitigation strategies, emergency response protocols, and the proper use of equipment and facilities. By adhering to stringent biosafety practices, laboratories can effectively mitigate the spread of infectious diseases, prevent laboratory accidents, and ensure the integrity and reliability of research and diagnostic activities.

ROUTINE DIAGNOSTICS

Routine diagnostics during pandemics are essential for early detection, surveillance, and management of infectious diseases. These diagnostics encompass various tests and screening methods aimed at identifying cases within populations, enabling early detection of even asymptomatic or mildly symptomatic cases. Regular diagnostic testing facilitates ongoing surveillance, tracking disease prevalence, transmission dynamics, and emerging hotspots within communities, informing evidence-based decision-making and resource allocation for pandemic response efforts. Diagnostic data guides policymakers in implementing or adjusting control measures such as lockdowns, travel restrictions, and vaccination campaigns based on the evolving epidemiological situation. Additionally, diagnostic testing is integral to contact tracing efforts, identifying and quarantining individuals in close contact with confirmed cases to prevent further transmission. It also supports resource allocation in healthcare systems, anticipating demand for medical care, hospital resources, and testing supplies, ensuring effective capacity planning and allocation of personnel and equipment to areas with the greatest need. Furthermore, routine diagnostics contribute to ongoing research and development efforts, driving innovation in testing technologies to enhance accuracy, efficiency, and accessibility, thereby strengthening pandemic preparedness and response strategies (Figure 1.3).

PUBLIC HEALTH INTERVENTIONS

Public health interventions serve as critical measures to combat pandemics, encompassing various strategies aimed at controlling the spread of infectious diseases. Testing and contact tracing efforts involve widespread

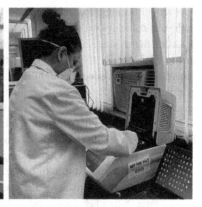

Figure 1.3 Adequate biosafety and routine diagnostics as key to combat pandemics like COVID-19. (Photo: Author.)

testing to identify cases and tracing contacts of infected individuals, thereby interrupting chains of transmission. Isolation and quarantine protocols are implemented to isolate confirmed cases and quarantine those exposed to the virus, effectively preventing further spread within communities. Public awareness campaigns play a pivotal role in educating the public about pandemic guidelines, symptom recognition, and the importance of seeking medical assistance when necessary, fostering community cooperation and adherence to preventive measures. Surveillance and monitoring efforts continuously track disease trends, hospital capacity, and other key metrics, providing essential data to inform decision-making and resource allocation for an effective pandemic response.

MEDICAL TREATMENTS

Medical treatments for pandemics such as COVID-19 have undergone significant evolution as researchers, and healthcare professionals have gained deeper insights into the virus and its impact on the human body. Supportive care forms the backbone of treatment for many patients, encompassing rest, hydration, and over-the-counter medications to manage symptoms. In severe cases, oxygen therapy becomes vital, administered through various methods, including nasal cannula or mechanical ventilation in intensive care units (ICUs). Proning, a technique positioning patients on their stomachs, aids in improving oxygenation and reducing reliance on mechanical ventilation. Fluid management is crucial, especially in severe cases where complications like pulmonary edema may arise. Drug therapies have also shown promise: antiviral medications such as remdesivir can reduce illness duration, while monoclonal antibodies like bamlanivimab and casirivimab/imdevimab are authorized for high-risk individuals (Bule M et al., 2019). Steroids like dexamethasone mitigate inflammation and reduce mortality in severe cases, and immunomodulators such as tocilizumab and baricitinib address cytokine storm syndrome. Additionally, anticoagulants help prevent or treat thrombotic complications. The development and deployment of vaccines, alongside ongoing clinical trials exploring new treatments, are crucial steps in the fight against COVID-19, underscoring the importance of collaborative efforts among healthcare providers, researchers, pharmaceutical companies, and regulatory agencies in delivering effective treatments to those in need.

FUTURE PROSPECTS

The 21st century has experienced a notable increase in severe infectious disease outbreaks, with the COVID-19 pandemic serving as a stark example of the devastating consequences on a global scale. Prior incidents, such as the 2003 SARS coronavirus outbreak, the 2009 swine flu pandemic, the 2012 MERS coronavirus outbreak, the 2013–2016 Ebola virus disease epidemic in West Africa, and the 2015 Zika virus disease epidemic, have all left lasting impacts by spreading across borders and affecting populations worldwide. This rise in outbreaks coincides with significant shifts in technology, demographics, and climate, including a doubling of airline flights since 2000, urbanization surpassing rural living since 2007, and escalating climate change threats (Baker et al., 2022). Changes in climate play a pivotal role in the emergence and resurgence of infectious diseases, alongside various human, biological, and ecological factors. Despite progress in sanitation and healthcare access, these global changes may exacerbate the risk of infectious disease outbreaks. Of particular concern is the role of climate change, which can influence the emergence and transmission of new pandemics through its effects on disease vectors, environmental conditions, and human susceptibility. Climatologists have observed a steady rise in global temperatures, projecting an unprecedented increase of 2.0°C by 2100. Of particular concern is the potential impact of these changes on the spread and prevalence of numerous serious infectious diseases. Mosquito-borne illnesses such as malaria, dengue, and viral encephalitis are notably sensitive to climatic shifts, with alterations in temperature directly influencing vector distribution, reproductive rates, biting behaviors, and pathogen incubation periods (Patz et al., 1996). Additionally, rising sea surface temperatures and sea levels associated with climate change heighten the risk of waterborne infectious diseases like cholera and shellfish poisoning. Indirectly, climate-related human migration and disruptions to healthcare infrastructure due to increased climate variability may exacerbate disease transmission dynamics. Furthermore, climate stress on agriculture could lead to malnutrition, while increased ultraviolet radiation exposure might impact human immune system function, potentially heightening susceptibility to infections. Given the anticipation of potential future pandemics, our research aims to propose an integrated healthcare system to enhance preparedness for forthcoming infectious outbreaks and pandemics. This system is envisioned to assist healthcare and disaster management officials, as well as policymakers, by facilitating data collection, sharing, and analysis (Nazayer et al., 2024). In this preview, the WHO has flagged **Disease X** (as a placeholder for future pandemics) as a major concern due to its epidemic potential and lack of medical countermeasures (Simpson et al., 2020). Starting with a convened meeting on November 18, 2022, the WHO has launched a comprehensive assembly of over 300 distinguished experts. The primary aim of this gathering is to thoroughly scrutinize available information on more than 25 virus families, bacteria, and a hypothetical pathogen (WHO).

CONCLUSION

In conclusion, understanding historical aspects of viral pandemics provides valuable insights into the emergence, spread, and impact of infectious diseases throughout history. By understanding past pandemics, we can better prepare for future outbreaks and improve our response strategies. Moving forward, leveraging advancements in science, technology, and global cooperation will be essential in mitigating the risks of future pandemics and safeguarding public health. The lessons learned from past experiences underscore the importance of proactive measures, robust healthcare systems, and international collaboration in addressing the challenges posed by viral pandemics.

REFERENCES

Andrews MA, Areekal B, Rajesh KR, Krishnan J, Suryakala R, Krishnan B, Muraly CP, Santhosh PV. First confirmed case of COVID-19 infection in India: A case report. *Indian J Med Res.* 2020;151(5):490–492. https://www.who.int/emergencies/diseases/novel-coronavirus-2019

Ashby B, Best A. Herd *immunity. Curr Biol.* 2021;31(4):R174–R177.

Ayouni I, Maatoug J, Dhouib W, Zammit N, Fredj SB, Ghammam R, Ghannem H. Effective public health measures to mitigate the spread of COVID-19: A systematic review. *BMC Public Health.* 2021;21(1):1015.

Baker RE, Mahmud AS, Miller IF, et al. Infectious disease in an era of global change. *Nat Rev Microbiol.* 2022;20:193–205.

Banik GR, Khandaker G, Rashid H. Middle East respiratory syndrome coronavirus "MERS-CoV": Current knowledge gaps. *Paediatr Respir Rev.* 2015;16(3):197–202.

Bule M, Khan F, Niaz K. Antivirals: Past, present and future. *Recent Adv Anim Virol.* 2019; 6:425–446.

Cheng SC, Chang YC, Fan Chiang YL, Chien YC, Cheng M, Yang CH, Huang CH, Hsu YN. First case of Coronavirus Disease 2019 (COVID-19) pneumonia in Taiwan. *J Formos Med Assoc.* 2020;119(3):747–751.

Cherry JD, Krogstad P. SARS: The first pandemic of the 21st century. *Pediatr Res.* 2004;56(1):1–5.

Chomel BB. *Encyclopedia of Microbiology.* 3rd ed. Elsevier Inc., University of California; Davis, CA: 2009. Zoonoses; pp. 820–829.

Cohen J. Swine flu pandemic. What's old is new: 1918 virus matches 2009 H1N1 strain. *Science.* 2010;327(5973):1563–1564. https://www.who.int/emergencies/situations/influenza-a-(h1n1)-outbreak

de Wit E, van Doremalen N, Falzarano D, et al. SARS and MERS: Recent insights into emerging coronaviruses. *Nat Rev Microbiol.* 2016;14:523–534.

Del Rio C, Guarner J. Ebola: Implications and perspectives. *Trans Am Clin Climatol Assoc.* 2015;126:93–112.

Duncan CJ, Scott S. What caused the Black Death? *Postgrad Med J.* 2005;81(955):315–20. doi: 10.1136/pgmj.2004.024075.

Falasca L, Agrati C, Petrosillo N, Di Caro A, Capobianchi MR, Ippolito G, Piacentini M. Molecular mechanisms of Ebola virus pathogenesis: Focus on cell death. *Cell Death Differ.* 2015;22(8):1250–1259.

Forni D, Cagliani R, Clerici M, Sironi M. Disease-causing human viruses: Novelty and legacy. *Trends Microbiol.* 2022;30(12):1232–1242.

Garten RJ, Davis CT, Russell CA, et al. Antigenic and genetic characteristics of swine-origin 2009 A(H1N1) influenza viruses circulating in humans. *Science.* 2009;325(5937):197–201.

Glatter KA, Finkelman P. History of the plague: An ancient pandemic for the age of COVID-19. *Am J Med.* 2021;134(2):176–181.

Grace D, Mutua F, Ochungo P, et al. Zoonoses Project 4. Report to the UK Department for International Development. International Livestock Research Institute; Nairobi, Kenya: 2012. Mapping of poverty and likely zoonoses hotspots.

Güner R, Hasanoğlu I, Aktaş F. COVID-19: Prevention and control measures in community. *Turk J Med Sci.* 2020;50(SI-1):571–577.

Highleyman L. First AIDS case in 1969. *BETA.* 1999;12(4):5. https://www.who.int/data/gho/data/themes/hiv-aids. Accessed on 23 December 2023.

Horton R. Offline: COVID-19 is not a pandemic. *Lancet.* 2020;396(10255):874.

Jester BJ, Uyeki TM, Jernigan DB. Fifty years of influenza A (H3N2) following the pandemic of 1968. *Am J Public Health.* 2020;110(5):669–676.

Kosik I, Yewdell JW. Influenza hemagglutinin and neuraminidase: Yin-Yang proteins coevolving to thwart immunity. *Viruses.* 2019;11(4):346.

Li F, Li W, Farzan M, Harrison SC. Structure of SARS coronavirus spike receptor-binding domain complexed with receptor. *Science.* 2005;309(5742):1864–1868.

Li YH, Hu CY, Wu NP, Yao HP, Li LJ. Molecular characteristics, functions, and related pathogenicity of MERS-CoV proteins. *Engineering.* 2019;5(5):940–947.

Madhav N, Oppenheim B, Gallivan M, et al. Pandemics: Risks, impacts, and mitigation. In: Jamison DT, Gelband H, Horton S, et al., editors. *Disease Control Priorities: Improving Health and Reducing Poverty.* 3rd ed. The International Bank for Reconstruction and Development/The World Bank; Washington (DC): 2017. Chapter 17.

McDermott A. Core concept: Herd immunity is an important-and often misunderstood-public health phenomenon. *Proc Natl Acad Sci U S A.* 2021;118(21):e2107692118.

McEvedy C. The bubonic plague. *Sci Am.* 1988;258(2):118–123. doi: 10.1038/scientificamerican0288-118.

Menon IG. The 1957 pandemic of influenza in India. *Bull World Health Organ.* 1959;20(2–3):199–224.

Michaud CM. Global burden of infectious diseases. In: Schaechter M, editor, *Encyclopedia of Microbiology:* 2009; pp. 444–454.

Milligan GN, Barrett AD. *Vaccinology: An Essential Guide.* Wiley Blackwell; Chichester, West Sussex: 2015; p. 310. ISBN 978-1-118-63652-7. OCLC 881386962.

Moghadami M. A narrative review of influenza: A seasonal and pandemic disease. *Iran J Med Sci*. 2017;42(1):2–13.

Mushegian AR. Are there 1031 virus particles on earth, or more, or fewer? *J Bacteriol*. 2020;202(9): e00052–e000520.

Nazayer M, Madanian S, Rasouli Panah H, Parry D. Pandemic management: Health data and public health surveillance. *Stud Health Technol Inform*. 2024. https://pubmed.ncbi.nlm.nih.gov/38269740/

Pal M, Berhanu G, Desalegn C, Kandi V. Severe Acute Respiratory Syndrome Coronavirus-2 (SARS-CoV-2): An update. *Cureus*. 2020;12(3):e7423.

Patz JA, Epstein PR, Burke TA, Balbus JM. Global climate change and emerging infectious diseases. *JAMA*. 1996;275(3):217–223.

Purohit N, Chugh Y, Bahuguna P, Prinja S. COVID-19 management: The vaccination drive in India. *Health Policy Technol*. 2022;11(2):100636.

Richard M, Knauf S, Lawrence P, Mather AE, Munster VJ, Müller MA, Smith D, Kuiken T. Factors determining human-to-human transmissibility of zoonotic pathogens via contact. *Curr Opin Virol*. 2017(22):7–12. https://www.who.int/news-room/fact-sheets/detail/zoonose. Accessed on 12 December 2023.

Riley LW. Differentiating epidemic from endemic or sporadic infectious disease occurrence. *Microbiol Spectr*. 2019;7(4):1–16.

Sabbatani S, Fiorino S. La peste antonina e il declino dell'Impero Romano. Ruolo della guerra partica e della guerra marcomannica tra il 164 e il 182 d.c. nella diffusione del contagio [The Antonine Plague and the decline of the Roman Empire]. *Infez Med*. 2009;17(4):261–275. Italian.

Schäfer JR, Kawaoka Y, Bean WJ, Süss J, Senne D, Webster RG. Origin of the pandemic 1957 H2 influenza A virus and the persistence of its possible progenitors in the avian reservoir. *Virology*. 1993;194(2):781–788.

Scire J, Nadeau S, Vaughan T, et al. Reproductive number of the COVID-19 epidemic in Switzerland with a focus on the Cantons of Basel-Stadt and Basel-Landschaft. *Swiss Med Wkly*. 2020;150:w20271.

Sharp PM, Hahn BH. Origins of HIV and the AIDS pandemic. Cold Spring Harb Perspect Med. 2011;1(1): a006841.

Simpson S, Kaufmann MC, Glozman V, Chakrabarti A. Disease X: Accelerating the development of medical countermeasures for the next pandemic. *Lancet Infect Dis*. 2020;20(5):e108–e115. https://www.who.int/news/item/21-11-2022-who-to-identify-pathogens-that-could-cause-future-outbreaks-and-pandemics

Snieszko SF. The effects of environmental stress on outbreaks of infectious diseases of fishes. *J Fish Biol*. 1974;6:197–208.

Strassburg MA. The global eradication of smallpox. *Am J Infect Control*. 1982;10(2):53–59.

Suzuki A. Smallpox and the epidemiological heritage of modern Japan: Towards a total history. *Med Hist*. 2011;55(3):313–318.

Tanaka S, Sugita S, Marui E. [A study on the 1946 smallpox epidemic in Japan and measures taken against it]. *Nihon Ishigaku Zasshi*. 2014;60(3):247–259. (in Japanese).

Taubenberger JK, Morens DM. The 1918 influenza pandemic and its legacy. *Cold Spring Harb Perspect Med*. 2020;10(10):a038695.

Thèves C, Biagini P, Crubézy E. The rediscovery of smallpox. *Clin Microbiol Infect*. 2014;20(3):210–218.

Turner BG, Summers MF. Structural biology of HIV. *J Mol Biol*. 1999;285(1):1–32.

van Seventer JM, Hochberg NS. Principles of infectious diseases: Transmission, diagnosis, prevention, and control. In: Stella R. Quah (ed.), *International Encyclopedia of Public Health*: 2017; pp. 22–39.

Wang Q, Chen H, Shi Y, Hughes AC, Liu WJ, Jiang J, Gao GF, Xue Y, Tong Y. Tracing the origins of SARS-CoV-2: Lessons learned from the past. *Cell Res*. 2021;31(11):1139–1141.

Woolhouse M, Scott F, Hudson Z, Howey R, Chase-Topping M. Human viruses: Discovery and emergence. *Philos Trans R Soc Lond B Biol Sci*. 2012;367(1604):2864–2871.

Yezli S, Otter JA. Minimum infective dose of the major human respiratory and enteric viruses transmitted through food and the environment. *Food Environ Virol*. 2011;3:1–30.

Yin F, Ji M, Yang Z, et al. Exploring the determinants of global vaccination campaigns to combat COVID-19. *Humanit Soc Sci Commun*. 2022;9:95. https://doi.org/10.1057/s41599-022-01106-7

2

Vitamin dysregulation and perturbations in viral diseases
Metabolism and biochemistry

RAKESH KAUSHIK AND MEGHNA GARG

GENERAL ROLE OF VITAMINS IN HUMANS AT MOLECULAR LEVEL

A vitamin is an organic molecule that is an essential micronutrient that an organism requires in small amounts for proper metabolism. Most vitamins are composed of groups of related molecules known as vitamers rather than single molecules. Vitamin E, for example, has eight vitamers: Four tocopherols and four tocotrienols. By including choline, some sources list fourteen vitamins, but major health organisations list 13 vitamins: vitamin A (all-trans-retinol, all-trans-retinyl-esters, all-trans-beta-carotene, and other provitamin A carotenoids), vitamin B1 (thiamine), vitamin B2 (riboflavin), vitamin B3 (niacin), vitamin B5 (pantothenic acid), vitamin B6 (pyridoxine), vitamin B7 (biotin), vitamin B9 (folic acid or folate), vitamin B12 (cobalamins), vitamin C (ascorbic acid), vitamin D (calciferols), vitamin E (tocopherols and tocotrienols), and vitamin K (phylloquinone and menaquinones).

Furthermore, these micronutrients are classified into two categories: water-soluble vitamins and fat-soluble vitamins. Water-soluble vitamins are not stored in our bodies and must be consumed on a daily basis. These are the B vitamins (B1, B2, B3, B5, B6, B8, B9, and B12) and vitamin C, which are found in varying amounts in plant and animal foods, as well as milk and its derivatives. In comparison, fat-soluble vitamins (A, D, E, and K) are absorbed along with dietary fats and accumulate in the liver and adipose tissue, with the exception of vitamin K, which is poorly stored in the tissues and thus requires a constant dietary supply. With the exception of vitamin D, which is synthesised by the human body, fat-soluble vitamins are mostly found in fruits and vegetables. Importantly, vitamins play a critical role in the regulation of numerous chemical reactions required for life. Vitamins, in general, act as coenzymes or prohormones by participating in a variety of human biological processes. It is well known that vitamins must be obtained through diet; however, in some cases, our bodies can synthesise them.

VITAMIN A

Vitamin A is an antioxidant vitamin, which means it can neutralise free radicals. As a result, adequate vitamin A intake can help to eliminate reactive oxygen species (ROS) and prevent the onset of diseases like heart failure and muscle damage.

The liver is important in vitamin A metabolism because it esterifies retinol to retinyl esters, which are then stored in stellate cells. Retinoids have a wide range of biological functions that are related to various vitamers: retinal regulates the function of rod photoreceptors, which are responsible for black-and-white vision in low-intensity light; retinal participates in glycoprotein synthesis, acting as a cofactor in the transport of mannose to the protein component; and retinoic acid acts as a transcriptional regulator. As a result, retinoic acid regulates cell differentiation and morphogenesis, is required for bone growth, ensures the integrity of skin and mucous membranes, is required for reproductive function (regulation of spermatogenesis and embryogenesis), and regulates granulocyte differentiation from myeloid stem cells [1]. It also plays a role in immunoregulatory processes. Vitamin A has been shown to regulate adaptive immunity and promote T-lymphocyte differentiation and

DOI: 10.1201/9781003435686-2

proliferation, particularly of regulatory T cells, memory B cells, and antibody-secreting plasma cells, especially those involved in IgA production [2]. Finally, antioxidative properties of carotenoids have been demonstrated.

VITAMIN D

Vitamin D is a fat-soluble vitamin that accumulates in the liver as a result of food consumption. Furthermore, when its use is required, it is released into circulation in small doses. Endogenous vitamin D production occurs when ultraviolet (UV) rays from sunlight strike the skin and initiate vitamin D synthesis [3]. In fact, there are two types of vitamin D: (i) Ergocalciferol, which is found in plant foods and (ii) cholecalciferol, which is synthesised by our bodies or found in plant foods. Cod liver oil, fatty fish (salmon, oysters, and shrimp), butter, egg yolk, mushrooms (the only vegetable source of vitamin D), and liver meat are all vitamin D-rich foods [4].

Furthermore, because vitamin D is a calcium metabolism regulator, it can help to calcify bones and prevent hypocalcaemic tetany (involuntary muscle contraction that causes cramps and spasms). Vitamin D also helps to keep calcium and phosphorus levels in the blood normal. Vitamin D also has other functions in the body, such as reducing inflammation and modulating processes like cell growth, neuromuscular and immune functions, and glucose metabolism. Vitamin D influences many genes that encode proteins that regulate cell proliferation, differentiation, and apoptosis. Furthermore, vitamin D receptors are found in many tissues, and some of them convert 25-hydroxyvitamin D (25 (OH) D) to 1,25 dihydroxyvitamin D (1,25 (OH) 2D) [5].

VITAMIN E

Vitamin E is a fat-soluble compound that has eight isoforms, four tocopherols (-, -, -, and -tocopherols), and four tocotrienols (-, -, -, and -tocotrienols). It is a lipid component of biological membranes. The various isoforms are incompatible, and -tocopherol is the most biologically active compound [6].

The primary source of vitamin E in the human diet varies depending on the isoform, with -tocopherol found primarily in nuts, almonds, hazelnuts, legumes, avocados, and sunflower seeds, with significant amounts also available in green leafy vegetables and fortified cereals.

Vitamin E protects cell membranes from free radical attack as a first-line defence against lipid peroxidation. Furthermore, vitamin E has the ability to inhibit lipid peroxidation by donating a hydrogen atom to peroxylipid radicals, making them less reactive. This redox reaction converts vitamin E into a reactively stable -tocopheroxy radical, which can react with vitamin C, glutathione, or coenzyme Q10 to reform -tocopherol. As a result, vitamin E is regarded as an important protective factor in all processes where negative effects of oxidative stress can occur, such as diseases like diabetes, cardiovascular and neurodegenerative diseases, and cancer, as well as physiological conditions like ageing or intense exercise [7].

VITAMIN K

Vitamin K, also known as naphthoquinone, is a fat-soluble vitamin that should not be consumed continuously through food. Vitamin K is actually stored in the body and used as needed. Vitamin K is required for the synthesis of prothrombin and other blood-clotting factors in the liver. It also helps to maintain the functionality of the proteins that keep bones healthy [8,9].

Vitamin K is classified into three groups based on its origin and function: vitamin K1 is of vegetable origin and participates in blood coagulation processes; vitamin K2 (also known as "menaquinone") is of bacterial origin, promotes intestinal microflora absorption, and is essential for bone health; and vitamin K3 ("water-soluble menadione") is of synthetic origin and is included in drugs that regulate blood coagulation processes. Vitamin K is mostly found in vegetables (tomatoes, spinach, cabbage, and turnip greens), but it is lacking in animal foods (with the exception of animal liver). Our intestinal bacterial flora also produces it. Furthermore, vitamin K is frequently regarded as a nutrient for improving heart health, lowering cancer risk, and increasing bone density [10].

VITAMIN B6

Vitamin B6 is one of the water-soluble vitamins, which cannot be stored in the body and must therefore be obtained through food. Vitamin B6 is a group of substances that includes pyridoxine, pyridoxal, pyridoxamine, and their 5'-phosphate esters [11]. The active coenzyme forms of vitamin B6 are pyridoxal 5'-phosphate (PLP) and pyridoxamine 5'-phosphate (PMP). PLP acts as a coenzyme in transamination reactions and in certain amino acid decarboxylation, deamination, and racemisation reactions. As a result, vitamin B6 is involved in amino acid, lipid, and carbohydrate metabolism. It also helps with cognitive development, immune function, and haemoglobin formation [12]. Fish, beef liver and other organs, potatoes, and other starch- and fruit-rich vegetables are the best sources of vitamin B6.

VITAMIN-INDUCED SIGNALLING IN INFECTIOUS VIRAL DISEASES

Nutrition is critical to the development and maintenance of the immune system. Nutritional deficiencies can impair the immune response and make you more susceptible to infections. A good nutritional status, on the other hand, can help to prevent the development of diseases and immune depression. Vitamins play a critical role in the regulation of many chemical reactions that occur in our bodies and are necessary for survival. They specifically aid in the supply of energy to the body and cell renewal, thereby preventing the onset of certain diseases such as neurocognitive disorders, muscle damage, and cardiac disorders and prevention of infectious viral diseases.

IMMUNOMODULATORY EFFECTS OF VITAMIN D

Vitamin D's pleiotropic role has been studied for decades, and there is solid evidence supporting an epidemiological link between low vitamin D level and a number of disorders. Recent research suggests that viral infections and vitamin D interact in a complicated way, including the induction of an antiviral state, functional immunoregulatory aspects, interaction with cellular and viral factors, induction of autophagy and death, and genetic and epigenetic modifications. Vitamin D receptors (VDRs) are found in nearly all cells of the body and at several locations throughout the genome. Vitamin D enters the body through dietary intake (approximately 20% of vitamin D3) or is synthesised by the skin (80%) from 7-dihydrocholesterol following seasonal exposure to type B ultraviolet (UVB). Sunlight's UVB rays convert 7-dihydrocholesterol to pre-vitamin D3, which then rapidly transforms to vitamin D3 (cholecalciferol). Following that, hydroxylation in the liver by enzymes cytochrome P450 2R1 (CYP2R1) and cytochrome P450 27 (CYP27A1) converts vitamin D3 to 25 hydroxyvitamin D3 (25(OH)D3). The renal 25-hydroxyvitamin D3-1 alpha-hydroxylase (CYP27B1) promotes the hydroxylation of 25-(OH)D3 in kidney tissue mitochondria, producing 1,25-(OH)2D3 (calcitriol), the active form of vitamin D3. Once activated, vitamin D attaches to distinct VDRs on different cells, where it can stimulate gene transcription, producing a hormone-like effect [13,14].

Vitamin D helps in activating the first line of defence, i.e., innate immunity during infections. The immune system's initial line of defence against alien intruders is innate immunity. It has been discovered that vitamin D is essential for monocyte activity. The enzyme CYP27B1 can establish a connection between monocytes and vitamin D [15]. Monocytes respond to foreign bodies by phagocytosing them and recognising them with toll-like receptors (TLRs) and other types of pattern recognition receptors. Some evidence suggests that when this event occurs, there is an increase in CYP27B1 activity [16]. This amplification suggests an increase in locally generated 1,25(OH)2D, which binds to the endogenous VDR and regulates gene expression in monocytes [17]. There is an increase in transcription of the gene that codes for the antibiotic protein LL37, which is the active form of the antimicrobial peptide cathelicidin [18]. An increase in the level of LL37 indicates improved monocyte function, demonstrating that vitamin D promotes monocyte activity and that vitamin D deficiency can have a deleterious impact on monocyte potency [18]. Calcitriol may also suppress inflammatory T-cell cytokines including IL-2 and IL-17, as well as TLRs on monocytes [19]. Calcitriol supplementation at high doses (1 g twice per day for 7 days) reduces the levels of the pro-inflammatory cytokine IL-6 released by peripheral mononuclear cells dramatically [20]. All of these actions are expected to combine to induce potential regulatory T cells, which are critical for regulating immunological responses and the development of autoreactivity.

Apart from directly combating microorganisms, monocytes and Antigen Presenting Cells (APCs), particularly dendritic cells (DC), are major targets for vitamin D's immunomodulatory actions. APCs are essential in the activation of the adaptive immune response because they are involved in the presentation of foreign antigens to T and B cells and are capable of regulating them via immunogenic or tolerogenic signals via cytokines and co-stimulatory molecule expression [19]. Several studies [19,21] have shown that vitamin D and its metabolites can modify the function and morphology of DC to create a more tolerogenic, immature state.

Not only innate immunity but also vitamin D plays a crucial role in regulating adaptive immunity of the host. Vitamin D inhibits the production of the cytokine IL-12, which is necessary for T-cell proliferation. Reduced IL-12 production leads to decreased IFN- and IL-2 production. IFN- and IL-2 are important cytokines for T-cell recruitment and proliferation [22]. Vitamin D also prevents T cells from developing into Th1 and Th17 cells, which are pro-inflammatory and have an effect on intracellular pathogens. Vitamin D has an inhibiting effect on B cells as well. While it has not been thoroughly investigated, it is known to suppress B-cell growth, differentiation to plasma cells, immunoglobulin production, and memory cell development, as well as promote apoptosis in B cells. VDR expression by T and B cells is very low in the resting state; however, upon activation and proliferation, there is a significant upregulation of VDR expression in T and B cells, allowing regulation of nearly 500 vitamin D responsive genes that affect the differentiation and proliferation of these cells [14,23,24].

IMMUNOMODULATORY EFFECTS OF VITAMIN C

Humans require vitamin C as a micronutrient. Although vitamin C is a powerful antioxidant that protects the body from both endogenous and exogenous oxidative stress, it is likely that its position as a cofactor for various biosynthetic and gene regulatory enzymes plays a significant part in its immune-modulating actions. Vitamin C promotes neutrophil recruitment to the site of infection, as well as phagocytosis, oxidant production, and microbial death. Simultaneously, it protects host tissue from severe damage by increasing neutrophil apoptosis and macrophage clearance and lowering neutrophil necrosis. As a result, it is clear that vitamin C is required for the immune system to build and sustain an appropriate response to infections while avoiding excessive host harm [25].

By increasing numerous immune cell functions, vitamin C appears to be capable of both preventing and treating respiratory and systemic infections. Vitamin C improves the skin's oxidant scavenging capacity and supports epithelial barrier function against infections, potentially protecting against environmental oxidative stress. Vitamin C accumulates in phagocytic cells like neutrophils and can improve chemotaxis, phagocytosis, ROS formation, and eventually microbial death. It is also required for apoptosis and macrophage clearance of wasted neutrophils from infection sites, reducing necrosis/NETosis, and possible tissue injury. The role of vitamin C in lymphocytes is less apparent; however, it has been found to improve B- and T-cell differentiation and proliferation, most likely due to gene-regulating actions. Vitamin C insufficiency impairs immunity and increases susceptibility to illnesses. In turn, infections have a considerable impact on vitamin C levels due to increased inflammation and metabolic needs. Furthermore, vitamin C supplementation appears to be capable of both preventing and treating respiratory and systemic infections. Prophylactic infection prevention necessitates appropriate, if not saturating, plasma levels of vitamin C (i.e., 100–200 mg/day), which optimises cell and tissue levels. Treatment of established infections, on the other hand, necessitates much greater (gram) dosages of the vitamin to compensate for the increased inflammatory response and metabolic requirement.

IMMUNOMODULATORY EFFECTS OF VITAMIN E

One of the most powerful nutrients known to affect immune function is vitamin E, a potent lipid-soluble antioxidant present in higher concentrations in immune cells than other cells in blood. Vitamin E insufficiency has been shown in animals and people to impede normal immune system activities, which can be addressed by vitamin E supplementation. Vitamin E influences T-cell activity directly by influencing T-cell membrane integrity, signal transduction, and cell division, as well as indirectly by influencing inflammatory mediators produced by other immune cells. Vitamin E's modulation of immune function has therapeutic implications because it impacts host susceptibility to infectious diseases such as respiratory infections, as well as allergy diseases such as asthma. Studies on the role of vitamin E in the immune system have usually focused on -tocopherol; however, new evidence suggests that other forms of vitamin E, such as other tocopherols and tocotrienols, may also have powerful immunomodulatory actions [23].

DIETARY ROLE OF VITAMIN SUPPLEMENTATION AND HOMEOSTASIS

Vitamin A is the generic name for a group of vitamers known as retinoids (retinol, retinal, and retinoic acid) that exhibit retinol's biological activity. Retinoids can be found in animal products (meat, fish, eggs, and derivatives). Carotenoids are found in green, yellow, and orange plants and act as provitamin A, producing at least one molecule of retinol via enzymatic hydrolysis. -Carotene, in particular, contributes to the orange colour of food and is commonly found in carrots and sweet potatoes.

The best food sources of B vitamins are as follows: B1: Organ meats (such as liver and kidney), eggs, nuts, seeds, whole grains, enriched grains, legumes, and peas; B2: Eggs, dairy products, organ meats, leafy greens, lean meats, legumes, and nuts; B3: Eggs, salt-water fish, poultry, enriched and whole grains, legumes, avocados, and potatoes; B5: Cabbage family vegetables (broccoli, cabbage, brussels sprouts, kale), eggs, organ meats, poultry, milk, mushrooms, legumes, lentils, white potatoes, sweet potatoes, and whole grains; B6: Meat and poultry, nuts, whole grains, avocado, bananas, and legumes; B7: Chocolate, egg yolks, legumes, nuts, dairy milk, organ meats, pork, and yeast; B9: Asparagus, broccoli and other cabbage family greens, leafy greens, beets, brewer's yeast, fortified grains, lentils, oranges, wheat germ, and peanuts; and B12: Eggs, dairy products, poultry, beef, pork, shellfish, organ meats, and fortified foods (such as fortified plant milks).

Vitamin C-rich foods includes citrus fruits, including oranges, lemons, limes, and grapefruit; semi-acidic fruits, such as mangoes, papayas, kiwi, pineapple, and cantaloupe; a variety of berries, including strawberries, blackberries, blueberries, cranberries, and raspberries; broccoli, brussels sprouts, cabbage, lettuce, turnip greens, spinach, collard greens and cauliflower, sweet potatoes, winter squash varieties, peppers, especially red and green varieties, tomatoes, and tomato products.

The very best source of vitamin D is sunshine, but plenty of foods contain trace amounts of vitamin D to support a well-rounded diet including fatty fish, such as tuna, mackerel, and salmon, egg yolks, beef liver, mushrooms, fortified milk, cheese made with fortified milk, and other fortified foods, such as orange juice, cereal, soy milk, and yogurt.

The best food sources of vitamin E are nuts, especially peanuts, almonds, and hazelnuts; seeds, especially pumpkin seeds and sunflower seeds; some vegetable oils, including wheat germ oil, safflower oil, sunflower oil, and soybean oil; leafy green vegetables; mangos; avocados; asparagus; red bell pepper; and fortified foods.

The dietary sources of vitamin K include eggs, poultry, pork, beef and organ meat, and leafy green vegetables, such as kale, spinach, arugula, Swiss chard, lettuce, collard greens and turnip greens, broccoli, cabbage, brussels sprouts, and cauliflower.

EVIDENCE OF VITAMIN PERTURBATIONS DURING VIRAL INFECTIONS

ROLE OF VITAMINS IN COVID-19 INFECTIONS

Recent research has revealed that vitamin D plays an important role in COVID-19-positive patients. Vitamin D promotes the expression and concentration of ACE2, MasR, and Ang- (1–7) and may protect against acute lung injury/acute respiratory distress syndrome. Vitamin C appears to be important in COVID-19-positive patients as well. Patients with acute respiratory infections, such as pneumonia or tuberculosis, have lower plasma vitamin C concentrations; however, vitamin C administration has been shown to reduce the severity and duration of pneumonia in elderly patients.

Vitamin C supplement has been shown to reduce the increase in pro-inflammatory cytokines such as IL-6 and TNF-, while also stimulating the production of anti-inflammatory cytokines such as interleukin-10 in COVID-19 patients.

EVIDENCE OF VITAMIN D PERTURBATIONS IN RHINOVIRUS, INFLUENZA VIRUS, RESPIRATORY SYNCYTIAL VIRUS, DENGUE VIRUS, HEPATITIS VIRUS, HIV, AND SARS-CoV2 VIRUS

Human rhinovirus (HRV) is the most common cause of upper respiratory tract infection (URI). Schneider et al. infected primary human bronchial epithelial cells (hBECs) with RV-16 after treating them with 25(OH)D or 1,25(OH)2D. They found that exposing the cells to vitamin D metabolites had no influence on RV-16 replication. However, both in the presence and absence of rhinovirus infection, incubation with vitamin D increased the release of pro-inflammatory chemokines CXCL8 and CXCL10. These chemokines are crucial in the recruitment of immune cells to the infection site, such as macrophages, neutrophils, and T cells, and so potentially operate as an effector mechanism in how vitamin D changes the antiviral response to HRV infection.

Previous research on adults infected with influenza viruses has revealed a higher incidence of hypovitaminosis due to factors affecting vitamin D metabolism, such as skin efficiency in synthesising vitamin D or reduced renal production of the active form of vitamin D, 1,25(OH)2D [26]. An observational study conducted in Norway between 1980 and 2000 discovered that low vitamin D levels in adults were associated with influenza-related mortality during the colder season [27]. An observational study on schoolchildren looked into the impact of vitamin D supplementation on influenza prevention. Some research studies have also looked into the effect of vitamin D on the efficiency of influenza virus immunisation. Some studies have found a correlation between vitamin D levels and vaccine efficacy.

RSV is the leading cause of bronchiolitis in one-year-old infants, with nearly all children showing serologic evidence of virus infection by the age of 2–3 years . Vitamin D deficiency has been linked to RSV vulnerability in studies, with low levels of 25(OH)D in cord blood plasma associated with RSV incidence in the first year of life. Furthermore, single-nucleotide polymorphisms in the VDR vitamin D-binding protein have been linked to a genetic predisposition to RSV bronchiolitis [28]. Vitamin D has the ability to reduce the inflammatory response to RSV infection while preserving the antiviral state and having no negative impact on viral load. This shows that vitamin D may play a role in lowering immunopathology, which is important in reducing disease severity and, as a result, morbidity and mortality from this frequent infection [29,30].

Dengue virus (DENV) is a highly endemic viral disease that is quickly becoming a global burden. Dengue fever is a tropical mosquito-borne disease caused by the dengue virus. Severe dengue sickness has been associated with high virus loads and an increase in pro-inflammatory cytokines, implying a malfunction in the mechanisms that govern cytokine production throughout the infection [31]. Controlling the release of pro-inflammatory cytokines by macrophages is an important step in preventing the course of dengue disease. A study found that differentiation of human monocyte-derived macrophages in the presence of vitamin D limits DENV infection via altering DENV binding to cells. Furthermore, they demonstrated that TNF-, IL-1, and IL-10 production was considerably lower in DENV-infected Monocyte-derived macrophages (MDMs) treated with vitamin D than in MDMs not treated with vitamin D. Vitamin D has been proven in studies to influence cytokine responses by indirect regulation of NF-B activity or direct manipulation of the VDR-dependent gene.

Infection with the hepatitis C virus is one of the primary causes of liver-related death. Vitamin D insufficiency is common in patients with liver disease. 25-OH vitamin D has numerous effects, including serving as an innate antiviral agent and providing an anti-fibrotic action, making it capable of delivering more than one benefit at the same time [32]. Furthermore, vitamin D is a potent immune modulator that has been linked to inflammatory responses as well as fibrosis caused by Hepatitis C Virus (HCV) infection. In individuals with chronic hepatitis C infection, vitamin insufficiency is prevalent. While it is known that HCV infection reduces vitamin D levels in individuals, it has been theorised that vitamin D levels before and during infection can alter disease outcome. One trial on immunocompetent patients with recurrent hepatitis C infection found that combining vitamin D with antiviral medication enhanced the likelihood of achieving a sustained viral response (SVR). Another study discovered that supplementing traditional Peg/Ribavirin (RBV) therapy with vitamin D significantly increases viral response in patients with HCV genotype 2–3. Low vitamin D levels have also been linked to severe fibrosis, in addition to a lower likelihood of SVR. The interaction of vitamin D with VDRs in fibroblasts has been shown to protect the cells from oxidative damage, impact fibroblast proliferation, gene expression, and migration and lower the inflammatory activity of liver stellate cells.

Some studies have shown that vitamin D levels in HIV patients can be utilised as prognostic indicators. A study on untreated African women found that low vitamin D levels were linked to a poor prognosis of the condition. According to one study, vitamin D insufficiency is linked to an increased risk of death. This could be linked to the role vitamin D plays in immune system regulation. Vitamin D stimulates the expression of antimicrobial peptides such as cathelicidin and defensin 2, and when vitamin D levels fall below 20 ng/mL, the cathelicidin response is not triggered, resulting in further immune system damage and an increase in opportunistic infections. Another study found that taking vitamin D supplements slowed the progression of the condition. As previously noted, vitamin D regulates CD4 function, and adequate levels of vitamin D limit viral entrance into cells and cytopathogenic consequences [22,33].

SARS-CoV-2 has been linked to severe respiratory illness and multi-organ failure. The majority of people infected with the virus show no symptoms, and just a small fraction of those who do show symptoms are badly impacted. This variation in symptoms can be due to individual differences in immunological response. According to clinical results, the virus enters the body and goes through a non-severe incubation period. If the host's immune system induces a specific immunological response during this stage, the virus can be eradicated before the disease worsens. The ability to remove the virus during the early stages of the disease is determined by the individual's immunological status (age, health condition, Human Leukocyte Antigen (HLA) type, ABO blood group, and others). If the virus is not eliminated by the host during the first stage, it will attack tissues with high ACE2 receptor expression, such as the lungs and kidneys. This will trigger inflammation mediated by the innate immune system, specifically pro-inflammatory cytokines such as (IL)-1B and IL-18 released by macrophages and type 1 T helper (Th1) released by immune cells, causing further organ damage and progressing the disease to the severe stage . Because vitamin D is essential for immune system regulation and is known to defend against viral infections, vitamin D levels are thought to correlate with the likelihood of SARS-CoV-2 infection and the result of COVID-19. COVID-19 is more prevalent in places that are higher in latitude and have little exposure to sunlight. This may be related to reduced vitamin D levels, which may enhance vulnerability to COVID-19. It has also been discovered that those who are classified as extremely vulnerable to COVID-19 have a vitamin D deficiency. Those suffering from chronic conditions such as diabetes and cardiovascular disease, as well as the obese and elderly, are considered high-risk groups for COVID-19 and, in most cases, are vitamin D deficient. SARS-CoV-2 primarily targets type II pneumocytes in the lungs, which are important in surfactant formation in the alveoli. When infected, these cells are killed by the intense immune reaction, which lowers surfactant synthesis in the lungs and frequently leads to problems in COVID-19 patients. Metabolites of 1,25-dihydroxy vitamin D have been shown in studies to increase surfactant formation in type II pneumocytes and to alleviate surface tension induced by surfactant depletion in COVID-19 patients. The majority of COVID-19 problems are triggered by a strong immunological reaction that destroys vital organs. Vitamin D has been reported to prevent excessive release of pro-inflammatory cytokines and chemokines by modifying macrophage activity and thereby reducing tissue damage.

EVIDENCE OF VITAMIN C PERTURBATIONS IN INFLUENZA INFECTIONS

Influenza is caused by a negative-stranded RNA virus infection that primarily attacks the lung epithelium and causes a severe local inflammatory response. Vitamin C is a necessary antioxidant as well as an enzymatic cofactor for physiological responses such as hormone production, collagen formation, and immunological potentiation [34]. It has long been investigated as a potential treatment for common colds, which can enhance vulnerability to influenza. Ritzel reported in 1961 that youngsters in a Swiss Alps ski school who were given vitamin C (1 g/day) had a lower incidence and duration of influenza infections. Pauling proposed in the early 1970s that vitamin C (1 g/day) could significantly reduce the frequency and severity of common cold episodes. Except for the common cold, vitamin C has been shown in human studies to relieve and prevent influenza symptoms when provided in megadoses before or after the onset of symptoms. In the human body, vitamin C operates quite differently than typical antiviral medications. Vitamin C is supposed to operate synergistically as a health supplement, interacting with the virus and the body to maintain the body in balance.

REFERENCES

1. Zasada M., Budzisz E. Retinoids: Active molecules influencing skin structure formation in cosmetic and dermatological treatments. *Postepy Dermatol. Alergol.* 2019;**36**:392–397. doi: 10.5114/ada.2019.87443.
2. Palace V.P., Khaper N., Qin Q., Singal P.K. Antioxidant potentials of vitamin A and carotenoids and their relevance to heart disease. *Free. Radic. Biol. Med.* 1999;**26**:746–761. doi: 10.1016/S0891-5849(98)00266-4.
3. Aranow C. Vitamin D and the immune system. *J. Investig. Med.* 2011;**59**:881–886. doi: 10.2310/JIM.0b01 3e31821b8755.
4. Umar M., Sastry K.S., Chouchane A.I. Role of vitamin D beyond the skeletal function: A review of the molecular and clinical studies. *Int. J. Mol. Sci.* 2018;**19**:1618. doi: 10.3390/ijms19061618.
5. Botelho J., Machado V., Proença L., Delgado A.S., Mendes J.J. Vitamin D deficiency and oral health: A comprehensive review. *Nutrients.* 2020;**12**:1471. doi: 10.3390/nu12051471.
6. Traber M.G., Stevens J.F. Vitamins C and E: Beneficial effects from a mechanistic perspective. *Free Radic. Biol. Med.* 2011;**51**:1000–1013. doi: 10.1016/j.freeradbiomed.2011.05.017.
7. Vardi M., Levy N.S., Levy A.P. Vitamin E in the prevention of cardiovascular disease: The importance of proper patient selection. *J. Lipid Res.* 2013;**54**:2307–2314. doi: 10.1194/jlr.R026641.
8. Van Ballegooijen A.J., Beulens J.W. The role of vitamin K status in cardiovascular health: Evidence from observational and clinical studies. *Curr. Nutr. Rep.* 2017;**6**:197–205. doi: 10.1007/s13668-017-0208-8.
9. Vermeer C. Vitamin K: The effect on health beyond coagulation-An overview. *Food Nutr. Res.* 2012;**56**: 5329. doi: 10.3402/fnr.v56i0.5329.
10. Yasin M., Butt M.S., Yasmin A., Bashir S. Chemical, antioxidant and sensory profiling of vitamin K-rich dietary sources. *J. Korean Soc. Appl. Biol. Chem.* 2014;**57**:153–160. doi: 10.1007/s13765-013-4235-x.
11. Kato N. Role of vitamin B6 in skin health and diseases. In: Preedy V.R., editor. *Handbook of Diet, Nutrition and the Skin.* Wageningen Academic Publishers; Wageningen, The Netherlands: 2012. Human Health Handbooks.
12. Young G. Leg cramps. *BMJ Clin. Evid.* 2015;**2015**:1113.
13. Siddiqui M., Mansansala J.S., Abdulrahman H.A., Nasrallah G.K., Smatti M.K., Younes N., Althani A.A., Yassine H.M. Immune modulatory effects of vitamin D on viral infections. *Nutrients.* 2020;**12**(9):2879. doi: 10.3390/nu12092879. PMID: 32967126; PMCID: PMC7551809.
14. Bikle D.D. Vitamin D metabolism, mechanism of action, and clinical applications. *Chem. Biol.* 2014;**21**:319–329. doi: 10.1016/j.chembiol.2013.12.016.
15. Adams J.S., Ren S., Liu P.T., Chun R., Lagishetty V., Gombart A.F., Borregaard N., Modlin R.L., Hewison M. Vitamin d-directed rheostatic regulation of monocyte antibacterial responses. *J. Immunol.* 2009;**182**:4289–4295. doi: 10.4049/jimmunol.0803736.
16. Holick M.F. Vitamin D: Important for prevention of osteoporosis, cardiovascular heart disease, type 1 diabetes, autoimmune diseases, and some cancers. *South. Med. J.* 2005;**98**:1024–1028. doi: 10.1097/01. SMJ.0000140865.32054.DB.
17. Heaney R.P. Vitamin D in health and disease. *Clin. J. Am. Soc. Nephrol.* 2008;**3**:1535–1541. doi: 10.2215/ CJN.01160308.

18. Baeke F., Takiishi T., Korf H., Gysemans C., Mathieu C. Vitamin D: Modulator of the immune system. *Curr. Opin. Pharmacol.* 2010;**10**:482–496. doi: 10.1016/j.coph.2010.04.001.

19. Banchereau J., Steinman R.M. Dendritic cells and the control of immunity. *Nature.* 1998;**392**:245–252. doi: 10.1038/32588.

20. Arruda E., Pitkäranta A., Witek T.J., A Doyle C., Hayden F.G. Frequency and natural history of rhinovirus infections in adults during autumn. *J. Clin. Microbiol.* 1997;**35**:2864–2868. doi: 10.1128/JCM.35.11.2864-2868.1997.

21. Belle A., Gizard E., Conroy G., Lopez A., Bouvier-Alias M., Rouanet S., Peyrin-Biroulet L., Pawlotsky J.M., Bronowicki J.-P. 25-OH vitamin D level has no impact on the efficacy of antiviral therapy in naïve genotype 1 HCV-infected patients. *United Eur. Gastroenterol. J.* 2017;**5**:69–75. doi: 10.1177/2050640616640157.

22. Sudfeld C.R., Wang M., Aboud S., Giovannucci E.L., Mugusi F.M., Fawzi W.W. Vitamin D and HIV progression among tanzanian adults initiating antiretroviral therapy. *PLOS ONE.* 2012;**7**:e40036. doi: 10.1371/journal.pone.0040036.

23. Lewis E.D., Meydani S.N., Wu D. Regulatory role of vitamin E in the immune system and inflammation. *IUBMB Life.* 2019 Apr;**71**(4):487–494. doi: 10.1002/iub.1976. Epub 2018 Nov 30. PMID: 30501009; PMCID: PMC7011499.

24. Cashman K.D., Dowling K.G., Škrabáková Z., González-Gross M., Valtueña J., De Henauw S., Moreno L., Damsgaard C.T., Michaelsen K.F., Mølgaard C., et al. Vitamin D deficiency in Europe: Pandemic? *Am. J. Clin. Nutr.* 2016;**103**:1033–1044. doi: 10.3945/ajcn.115.120873.

25. Carr A.C., Maggini S. Vitamin C and immune function. *Nutrients.* 2017;**9**(11):1211. doi: 10.3390/nu9111211. PMID: 29099763; PMCID: PMC5707683.

26. Stumpf W., Sar M., Reid F., Tanaka Y., DeLuca H. Target cells for 1,25-dihydroxyvitamin D3 in intestinal tract, stomach, kidney, skin, pituitary, and parathyroid. *Science.* 1979;**206**:1188–1190. doi: 10.1126/science.505004.

27. Cannell J., Vieth R., Umhau J.C., Holick M.F., Grant W.B., Madronich S., Garland C.F., Giovannucci E. Epidemic influenza and vitamin D. *Epidemiol. Infect.* 2006;**134**:1129–1140. doi: 10.1017/S0950268806007175.

28. Beigelman A., Castro M., Schweiger T.L., Wilson B.S., Zheng J., Yin-DeClue H., Sajol G., Giri T., Sierra O.L., Isaacson-Schmid M., et al. Vitamin D levels are unrelated to the severity of respiratory syncytial virus bronchiolitis among hospitalized infants. *J. Pediatr. Infect. Dis. Soc.* 2015;**4**:182–188. doi: 10.1093/jpids/piu042.

29. Currie S.M., Findlay E.G., McHugh B.J., Mackellar A., Man T., Macmillan D., Wang H., Fitch P.M., Schwarze J., Davidson D.J. The human cathelicidin LL-37 has antiviral activity against respiratory syncytial virus. *PLOS ONE.* 2013;**8**:e73659. doi: 10.1371/journal.pone.0073659.

30. Kota S., Sabbah A., Chang T.H., Harnack R., Xiang Y., Meng X., Bose S. Role of human β-defensin-2 during tumor necrosis factor-α/NF-kappaB-mediated Innate antiviral response against human respiratory syncytial virus. *J. Biol. Chem.* 2008;**283**:22417–22429. doi: 10.1074/jbc.M710415200.

31. Pang T., Cardosa M.J., Guzman M.G. Of cascades and perfect storms: The immunopathogenesis of dengue haemorrhagic fever-dengue shock syndrome (DHF/DSS). *Immunol. Cell Biol.* 2007;**85**:43–45. doi: 10.1038/sj.icb.7100008.

32. Assy N., Mouch A. Vitamin D improves viral response in hepatitis C genotype 2-3 naïve patients. *World J. Gastroenterol.* 2012;**18**:800–805. doi: 10.3748/wjg.v18.i8.800.

33. Viard J.-P., Souberbielle J.-C., Kirk O., Reekie J., Knysz B., Losso M., Gatell J., Pedersen C., Bogner J.R., Lundgren J., et al. Vitamin D and clinical disease progression in HIV infection: Results from the EuroSIDA study. *AIDS.* 2011;**25**:1305–1315. doi: 10.1097/QAD.0b013e328347f6f7.

34. Bae M, Kim H. The role of vitamin C, vitamin D, and selenium in immune system against COVID-19. *Molecules.* 2020;**25**(22):5346. doi: 10.3390/molecules25225346.

3

The significance of water-soluble vitamins in SARS and related infections

SHILPA MAHAJAN, RAKESH KAUSHIK, AND VIVEK DHAR DWIVEDI

INTRODUCTION

There is a significant impact of respiratory diseases on global morbidity over the last 15 years. Specifically, respiratory diseases have consistently held the top rank, comprising approximately 24%–25% of general morbidity worldwide (Wang et al., 2019). It is noted that acute respiratory infections of the respiratory tract stand out as the most common cause of both cases and prolonged temporary disability compared to other acute pathologies. It highlights the extensive and persistent health hurdles presented by respiratory illnesses, notably acute respiratory infections, worldwide. Annually, more than 200 million respiratory tract infections occur globally, excluding pandemic periods, with less than one-third of these infections reported to healthcare providers (Calderero et al., 2022). The majority of reported cases are handled by healthcare providers within the outpatient network, predominantly in primary care settings. Respiratory tract infections represent over 10% of all visits to primary care offices. These infections are broadly categorized into upper and lower respiratory tract infections, with acute bronchitis and pneumonia classified as lower respiratory tract infections, while severe acute respiratory syndrome (SARS), influenza, pharyngitis, sinusitis, otitis media, and bronchitis cause upper respiratory tract infections. Furthermore, it underscores the significant repercussions of complications, particularly in the context of pneumonia and influenza, on morbidity, economic ramifications, and mortality, especially among the elderly. Such complications can result in severe outcomes, potentially culminating in fatalities. Pneumonia and influenza, particularly when occurring together, are highlighted as the fourth leading cause of death in individuals aged 65 and older (WHO, 2019).

In the late 1920s, the first documented cases of a coronavirus infection in animals were observed. This occurred when a novel acute respiratory infection affected domesticated chickens in North America. Subsequently, in the 21st century, three significant outbreaks of human coronavirus diseases emerged: SARS, Middle East respiratory syndrome (MERS), and coronavirus disease 2019 (COVID-19). These outbreaks were attributed to distinct coronaviruses (CoVs). The common belief is that these viruses originated in bats and were transmitted to humans through intermediate hosts (Chen et al., 2020). Coronaviruses are described as having crown-like particles. The particles have spikes protruding from their surfaces. Coronaviruses are classified as enveloped viruses. They have single-stranded, positive-sense RNA (+ssRNA) genomes. The genome size of coronaviruses is specified to be approximately 26–32 kilobases (kb). This genome size is noted as the largest known among RNA viruses (Su et al., 2016). These viruses belong to the subfamily Orthocoronavirinae in the Coronaviridae family and the order Nidovirales. Before the year 2002, coronaviruses were considered to cause only mild diseases in humans. Specifically, two human coronaviruses, HCoV-OC43 and HCoV-229E, were associated with non-hospitalized respiratory infections. The outbreak of SARS-CoV-1 in 2002 changed the view regarding coronaviruses' potential to cause severe diseases in humans. This event marked a significant shift in understanding, indicating that coronaviruses could lead to more serious and severe respiratory illnesses (Cui et al., 2019). The SARS-CoV-1 outbreak lasted for 9 months. During the outbreak, 8098 people were infected with SARS-CoV-1. There were 774 reported deaths associated with the infection. The SARS-CoV-1 outbreak is characterized as causing the first global pandemic in the 21st century (Rota et al., 2003).

DOI: 10.1201/9781003435686-3

Seventeen years later, another coronavirus-related respiratory disease outbreak occurred, referring to COVID-19 (Huang et al., 2020; Zhou et al., 2020). The disease agent responsible for COVID-19 is identified as a novel SARS-related coronavirus, named SARS-CoV-2 (Zhou et al., 2020). COVID-19 has spread to nearly all countries in the world. It has caused a serious public health crisis on a global scale. There have been more than 100 million reported cases of COVID-19 infection globally. There have been 2.5 million reported deaths attributed to COVID-19 as of April 24, 2021. SARS-CoV-2 shares 79.6% genome sequence identity with SARS-CoV-1. Both viruses exhibit many biological features in common. Both viruses use the same cell entry receptor, angiotensin-converting enzyme 2 (ACE2), for entry into human cells. Both viruses cause severe pneumonia and systemic inflammatory diseases in humans (Zhou et al., 2020; Hu et al., 2021). SARS-CoV-2 is noted to be more transmissible compared to SARS-CoV-1. SARS-CoV-2 exhibits a high viral titer in the respiratory system early before the onset of symptoms. In contrast, high SARS-CoV-1 viral load is typically detected in more severe patients (Wölfel et al., 2020). The difference in transmissibility is suggested to partially explain why there are significantly more COVID-19 patients compared to SARS cases.

There are seven coronaviruses that infect humans; among these, HCoV-229E, HCoV-NL63, HCoV-OC43, and HCoV-HKU1 usually cause mild respiratory symptoms similar to the common cold (van der Hoek, 2007), whereas SARS-CoV, MERS-CoV, and SARS-CoV-2 are severe coronaviruses. SARS-CoV (severe acute respiratory syndrome coronavirus) emerged in 2002–2003 (Zhong et al., 2003) and caused a global outbreak with a relatively high mortality rate. MERS-CoV (Middle East respiratory syndrome coronavirus) emerged in 2012 (Zaki et al., 2012) and also led to severe respiratory illness, particularly in the Middle East. SARS-CoV-2, the virus responsible for COVID-19, emerged in late 2019 and has caused a worldwide pandemic. It is distinct from SARS-CoV and MERS-CoV but belongs to the same betacoronavirus group. Coronaviruses (CoVs) are known to have a relatively moderate to high mutation rate compared to some other single-stranded RNA (ssRNA) viruses (Su et al., 2016). The high mutation rate is a characteristic of RNA viruses in general, and it is due to the lack of a proofreading mechanism during the replication of their genomes. The RNA-dependent RNA polymerase (RdRp) enzyme, responsible for copying the viral RNA during replication, lacks proofreading activity in many RA viruses, including coronaviruses. This means that errors or mutations can occur more frequently during the viral genome replication process.

The high mutation rate of coronaviruses carries several significant implications. Firstly, it fosters genetic diversity, resulting in a wide array of virus variants even within a single infected individual. This diversity facilitates the virus's adaptation to various environments and hosts. Secondly, the rapid mutation rate grants RNA viruses, including coronaviruses, substantial evolutionary potential. This can influence the virus's capacity to evade host immune responses, develop resistance to antiviral medications, or even adapt to new host species. Thirdly, the genetic diversity poses challenges for vaccine development, as changes in the virus's genetic makeup may affect its antigenic properties. This necessitates regular updates to vaccines, akin to those required for influenza, which targets a swiftly mutating virus. It is crucial to recognize that while coronaviruses exhibit a moderate to high mutation rate, not all mutations lead to significant alterations in the virus. Many mutations may be neutral or even harmful, with only a subset conferring a selective advantage. Monitoring the genetic diversity of coronaviruses is thus vital for comprehending their evolution and guiding public health strategies, including vaccine development and the design of antiviral drugs. Furthermore, understanding the similarities and differences between coronaviruses such as SARS-CoV (severe acute respiratory syndrome coronavirus), SARS-CoV-2, and MERS-CoV (Middle East respiratory syndrome coronavirus) regarding their invasion mechanisms, interaction with cellular receptors, and modulation of the host cell immune response is imperative for elucidating the pathogenesis of these diseases and informing effective control measures.

SARS-CoV and SARS-CoV-2 both use angiotensin-converting enzyme 2 (ACE2) as their cellular receptor for entry into host cells. SARS-CoV-2, in addition to ACE2, has been reported to interact with CD147. MERS-CoV uses dipeptidyl peptidase 4 (DPP4 or CD26) as its cellular receptor. SARS-CoV and SARS-CoV-2 primarily infect the respiratory system, targeting cells expressing ACE2, including those in the lungs. MERS-CoV has a different tissue tropism, infecting cells in the respiratory and gastrointestinal tracts.

SARS-CoV and SARS-CoV-2 have been reported to impact the renin–angiotensin system (RAS) by suppressing ACE2. ACE2 is involved in regulating the RAS, and its downregulation can lead to dysregulation of this system, potentially contributing to the onset of symptoms observed in SARS and COVID-19. Understanding the interactions between these viruses and host cells, as well as the impact on the host immune system and physiological pathways, is crucial for developing effective treatments and interventions. It's worth noting that research in this field is ongoing, and new findings may further refine our understanding of these viruses and their interactions with the host. It emphasizes a critical aspect of public health and global preparedness in the face of epidemic outbreaks. Indeed, the 21st century has witnessed several significant epidemic events, such as the SARS outbreak in 2002–2003, the H1N1 influenza pandemic in 2009, and, most prominently, the COVID-19 pandemic caused by SARS-CoV-2.

Epidemic outbreaks have far-reaching consequences, affecting not only public health but also the economy, psychology, and human behavior. The interconnectedness of these aspects underscores the need for comprehensive responses. Future preparedness should focus on developing and implementing precautionary measures that guide individuals and communities in taking effective emergency actions. This includes strategies for containing the spread of the virus, protecting vulnerable populations, and maintaining social stability.

Ongoing research on coronaviruses and other epidemic-causing agents is crucial for advancing our comprehension of these pathogens. This knowledge serves as the foundation for crafting more robust responses, medical treatments, vaccines, and strategies to manage personal anxiety during outbreaks. International collaboration and information sharing are paramount for a unified response to epidemics, encompassing data exchange, joint research initiatives, and the establishment of standardized protocols for preparedness and response. Recognizing the interconnected nature of health, efforts should adopt a holistic approach that addresses both physical and mental well-being, integrating mental health support into public health responses.

As we move forward, global efforts in research, healthcare infrastructure, and public health policies will play a pivotal role in enhancing our ability to respond effectively to emerging infectious diseases and mitigating their impact on societies worldwide. Future research should focus deeper into understanding the unique characteristics of each virus, including their genetic makeup, replication mechanisms, and interactions with host cells, strategies to evade immune system. Comparative studies can identify commonalities and distinctions, providing insights into the evolution and pathogenicity of these viruses. Research efforts should explore new treatment modalities, vaccine platforms, and combination therapies to enhance efficacy and reduce the risk of resistance. Overall, future research directions should be comprehensive, multidisciplinary, and focused on addressing the specific challenges posed by each virus, ultimately contributing to improved preparedness and response capabilities in the face of emerging viral threats.

WATER-SOLUBLE VITAMINS: AN OVERVIEW

Vitamins constitute a diverse group of organic nutrients essential in small quantities for various biochemical reactions crucial for an organism's growth, survival, and reproduction. Typically, the body cannot produce these compounds, necessitating their intake through the diet. The primary role of vitamins is often as coenzymes, facilitating enzymatic reactions. The exploration of vitamins originated in early 20th-century experiments by Hopkins, who observed stunted growth in rats on a defined diet until the addition of a small amount of milk restored normal development. This led to the concept of "accessory growth factors" present in small amounts in certain foods. In 1912, Funk coined the term "vitamine," emphasizing its chemical nature. While subsequent factors were not amines, the term vitamin persists. Pronunciation varies, with both "vitamin" and "veitamin" considered acceptable. In the early 20th century, vitamin deficiencies were prevalent, but by the 21st century, they became rare in developed countries. However, vitamin A deficiency persists globally, and subclinical deficiencies of B2 and B6 are documented. Vulnerable populations, such as refugees, face heightened risks due to the absence of micronutrient fortification in emergency rations. Different chemical compounds that exhibit the same biological activity are collectively referred to as vitamers. In cases where multiple compounds share biological activity, alongside individual names, an approved generic descriptor is designated for use across all related compounds that demonstrate the same biological activity. This generic descriptor helps streamline the communication and understanding of the shared functional properties among these compounds.

The realization that milk contained more than one accessory food factor led to the naming of these factors as A (lipid soluble, found in the cream) and B (water soluble, found in the whey). This classification into fat-soluble and water-soluble vitamins persists, even though there is little chemical or nutritional justification for it, except for some similarities in the dietary sources of these vitamins. Over time, water-soluble derivatives of vitamins A and K, as well as fat-soluble derivatives of several B vitamins and vitamin C, have been synthesized for therapeutic use and as additives in food. There are 13 vitamins found in the human body; among these, 4 are fat soluble and 9 are water soluble. The former are soluble in fat and can be absorbed through gastrointestinal tract, while the latter are soluble in water and can be excreted out through urine and cannot be stored in body.

In general, the majority of vitamins are acquired through dietary intake, but certain vitamins can be obtained through alternative means. For instance, microorganisms in the gut flora produce vitamin K and biotin, while skin cells synthesize one form of vitamin D when exposed to specific ultraviolet light from sunlight. Some vitamins can be internally produced by the human body from consumed precursors, such as the synthesis of vitamin A from beta-carotene and niacin from the amino acid tryptophan. The ability to synthesize vitamin C varies among species. Notably, vitamin B12 is unique in that it is not available from plant sources. The Food Fortification Initiative identifies countries with mandatory fortification programs for vitamins like folic acid, niacin, vitamin A, and vitamins B1, B2, and B12.

The storage of vitamins in the body varies from person to person. Both deficient and excessive intake of vitamins may cause several disorders. Vitamins, in general, do not possess plastic properties. However, an exception exists with vitamin F, which exhibits characteristics reminiscent of plastic materials. Typically, vitamins do not serve as an energy source for the body. An exception to this rule is vitamin F, which plays a unique role in providing energy-related functions. Even in modest amounts, vitamins play a crucial role in supporting all essential bodily functions, demonstrating their biological activity across various vital processes. Vitamins exert their impact on biochemical processes throughout all tissues and organs, showcasing their non-specificity to any particular organ and their widespread influence across the body. The definition of a compound as a vitamin necessitates its identification as a dietary essential. Its absence from the diet should lead to a distinct deficiency disease, and reintroduction must either cure or prevent that deficiency ailment. Merely demonstrating pharmacological actions or the ability to treat a disease, even if the compound occurs naturally in foods, does not warrant classification as a vitamin. Similarly, establishing a physiological function as a coenzyme or hormone is insufficient for vitamin classification; it is crucial to prove that endogenous synthesis is inadequate to meet physiological needs in the absence of dietary intake.

Vitamins are divided into two categories based on their absorption and storage characteristics. Water-soluble vitamins dissolve in water upon entering the body, preventing humans from storing excess amounts for future use. There are nine water-soluble vitamins: B vitamins (folate, thiamine, riboflavin, niacin, pantothenic acid, biotin, vitamin B6, and vitamin B12) and vitamin C. Deficiency in any of these vitamins can lead to clinical syndromes with potential severe morbidity and mortality.

SOURCES, ABSORPTION, AND BIOAVAILABILITY OF WATER-SOLUBLE VITAMINS

The focus here is on recent advancements in comprehending the cellular and molecular mechanisms governing the regulatory processes of intestinal absorption for various vitamins. The specific focus encompasses water-soluble vitamins such as folic acid, cobalamin (vitamin B12), biotin, pantothenic acid, and thiamine (vitamin B1), along with the lipid-soluble vitamin A. Regarding folate, notable progress has been achieved, including the elucidation of the digestive conversion of dietary folate polyglutamates to monoglutamates through the identification of the responsible enzyme. Additionally, advancements involve the discovery of the cDNA responsible for the intestinal folate transporter, clarification of intracellular mechanisms governing small intestinal folate uptake, and the identification and characterization of a specific pH-dependent, carrier-mediated system for folate uptake at the luminal (apical) membrane of human colonocytes.

Studies on cobalamin have concentrated on the cellular and molecular characterization of the intrinsic factor and its receptor. Investigations into biotin transport in the small intestine have revealed a shared uptake process with another water-soluble vitamin, pantothenic acid. Furthermore, a Na^+-dependent, carrier-mediated biotin uptake system, also shared with pantothenic acid, has been identified at the apical membrane of human colonocytes. This carrier is presumed to be responsible for absorbing bacterially synthesized biotin and pantothenic acid in the large intestine. Preliminary studies have reported the cloning of a biotin transporter from the small intestine. Concerning thiamine intestinal transport, research indicates that thiamine uptake by small intestinal biopsy specimens occurs through a carrier-mediated, Na-independent mechanism, which appears to be up-regulated in thiamine deficiency. Investigations into vitamin A intestinal absorption reveal the existence of a receptor-mediated mechanism for the uptake of retinol bound to retinol-binding protein in the small intestine of suckling rats. Another study indicates that retinoic acid increases the mRNA level of cellular retinol-binding protein II and the rate of retinol uptake by Caco-2 intestinal epithelial cells. The study suggests that retinoids may play a role in regulating vitamin A intestinal absorption.

The biological availability of a nutrient refers to the proportion of that nutrient present in a food that can be utilized by the body. This availability is determined by several factors, including the extent of nutrient digestion, the absorption of digestion products, and the metabolism of these products. Various factors come into play, influencing digestion, absorption, metabolism, and, consequently, the overall biological availability. These factors encompass the physical characteristics of the food matrix (for example, nutrients enclosed within intact cells of plant foods, where the plant cell wall remains undigested), the chemical nature of the vitamin in the food, and the presence of inhibitors, which may be inherent in the food, consumed with the food, or taken as drugs or medications (Bates and Heseker, 1994; Ball, 2013). Numerous vitamins undergo absorption through active transport, a saturable process. Consequently, the percentage of absorption tends to decrease as intake levels increase. Furthermore, several water-soluble vitamins are bound to proteins in foods, and their release may necessitate either the action of gastric acid (as observed in the case of vitamin B12) or specific enzymatic hydrolysis. For example, the action of conjugates is required to hydrolyze folate conjugates, and the hydrolysis of biocytin is needed to release biotin. The state of body reserves of the vitamin can influence its absorption by affecting the synthesis of binding and transport proteins. Additionally, it can impact the extent to which the vitamin is metabolized after uptake into the intestinal mucosa. For instance, the oxidative cleavage of carotene to retinaldehyde is regulated by vitamin A status.

Certain vitamins in foods exist in chemical forms that resist enzymatic hydrolysis during digestion, although they may be released during food preparation for analysis. For instance, much of the vitamin B6 in plant foods is present as pyridoxine glycosides, which are only partially available and may antagonize the metabolism of free pyridoxine (Gregory, 1998). Excessive heating can lead to the nonenzymic formation of pyridoxyllysine in foods, rendering both the vitamin and lysine unavailable. Additionally, the majority of niacin in cereals is present as niacytin (nicotinoyl-glucose esters in oligosaccharides and non-starch polysaccharides), which is only partially hydrolyzed by gastric acid. Also, sometimes, drugs and compounds naturally present in foods have the potential to compete with vitamins for absorption and therefore influence their bioavailability. For example, chlorpromazine, tricyclic antidepressants, and certain antimalarial drugs can inhibit the intestinal transport and metabolism of riboflavin. Carotenoids lacking vitamin A activity may compete with β-carotene for intestinal absorption and metabolism. Moreover, alcohol has been observed to inhibit the active transport of thiamin across the intestinal mucosa. The bioavailability of nutrients, referring to the portion of nutrients released from digested food that is accessible for absorption into the gut and subsequently available for host cell metabolism, is a critical factor. Various methods are commonly employed to assess bioavailability, including in vitro approaches such as simulated gastrointestinal digestion, Caco-2 cell studies, and examination of cell membranes. Additionally, ex vivo methods involve studying gastrointestinal organs under controlled laboratory conditions. Finally, in vivo studies, conducted in both human and animal subjects, provide insights into the actual absorption and utilization of nutrients (Barba et al., 2017). Understanding the bioavailability of nutrients is essential for comprehending their impact on host cell metabolism and designing strategies to optimize nutrient absorption for overall health. The bioavailability of nutrients can vary significantly between macronutrients and micronutrients, as noted by Carbonell-Capella et al. (2014). Various external factors influence nutrient absorption, including the structure of food matrices, the form of the nutrient, combinations with other nutrients, and the quantity of non-nutrient components. Additionally, internal factors such as age, gender, physiological status, and nutritional status also play a crucial role in governing the absorption of nutrients. These multifaceted considerations underscore the complexity of nutrient bioavailability, emphasizing the need for a comprehensive understanding of both external and internal factors to optimize nutritional outcomes and support overall health. There are several other factors like cooking methods, storage methods, and the way of interactions with other nutrients that can influence the bioavailability of water-soluble vitamins. B vitamins also show interdependence on each other; i.e., deficiency of one vitamin affects the functioning of other vitamins. The individual genetic variation can influence the absorption, metabolism, and utilization of water-soluble vitamins (Aleksandrova et al., 2016).

ROLE OF VITAMIN C IN IMMUNE FUNCTIONS

Dietary antioxidants play a crucial role in safeguarding the airways from the detrimental impacts of oxidative stress, a characteristic feature of respiratory diseases (Wood et al., 2003). Oxidative stress, triggered by reactive oxygen species (ROS), is produced in the lungs due to various exposures like air pollution, airborne irritants, and typical airway inflammatory cell responses (Kelly, 2005). Elevated ROS levels can escalate inflammation in the airways through the activation of NF-κB and the expression of pro-inflammatory mediators (Rahman, 2003). Fruits, vegetables, nuts, vegetable oils, cocoa, red wine, and green tea are rich sources of antioxidants, including vitamin C, vitamin E, flavonoids, and carotenoids. The presence of these antioxidants in the diet may exert positive effects on respiratory health, influencing outcomes from maternal diet effects on the fetus to intake in children, adults, and pregnant women with asthma, as well as adults with Chronic Obstructive Pulmonary Disease (COPD).

Observational studies in children have shown that the consumption of fruits, rich in vitamin C, is associated with reduced wheezing (Forastiere et al., 2000). Some epidemiological studies demonstrated a positive association between vitamin C intake and lung function (Schwartz and Weiss, 1994), while others did not find such associations (Butland et al., 2000; Shaheen et al., 2001). Despite observational data linking vitamin C to lung health, supplementation with vitamin C has not been proven to reduce the risk of asthma (Shaheen et al., 2001). This lack of efficacy may be attributed to the interdependence of nutrients in foods, suggesting that supplementing with isolated nutrients might not be as effective as obtaining them from whole foods.

Vitamin C plays several important roles in supporting the immune system and maintaining overall health. It plays a crucial role in collagen synthesis, which is essential for maintaining the integrity of epithelial barriers. Epithelial barriers serve as a physical defense against pathogens and other harmful agents. Vitamin C is involved in stimulating the production, functioning, and movement of leukocytes (white blood cells) in the innate immune system. This includes neutrophils, lymphocytes, and phagocytes, which are crucial for the body's initial response to infections. Vitamin C increases the serum levels of complementary proteins like neutrophils and lymphocytes. These proteins are key components of the immune system and play vital roles in defending the body against infections. Natural killer (NK) cells, which are a type of lymphocyte involved in the innate immune response, are known to be affected by the activity of vitamin C. Vitamin C stimulates macrophages, which are specialized immune cells. Macrophages play a role in various immune functions, including chemotaxis (movement toward a chemical signal), apoptosis (programmed cell death), and the removal of spent neutrophils from sites of infection. Vitamin C exhibits antimicrobial effects, contributing to the body's defense against pathogens. Vitamin C acts as a protective antioxidant, helping to neutralize reactive oxygen species (ROS) and reactive nitrogen species (RNS) that can be generated during the immune system's destruction of pathogens. This antioxidant function helps reduce oxidative stress and inflammation. Thus, vitamin C serves as a potent antioxidant, playing a key role in neutralizing harmful free radicals in the body.

It is quite evident that vitamin C is a multifaceted nutrient that plays a vital role in supporting various aspects of the immune system and overall health. Consuming an adequate amount of vitamin C through a balanced diet or supplementation can contribute to a robust immune response and help maintain the body's defense against infections. Vitamin C can increase the serum levels of B and T antibodies in the adaptive immune system and play an active role in the differentiation and proliferation of lymphocytes (Wintergerst et al., 2006; Haryanto et al., 2015; Carr and Maggini, 2017).

While certain studies suggest potential benefits of vitamin C in respiratory infections, results can vary, and the overall evidence is not universally conclusive. The effectiveness of vitamin C may depend on factors such as dosage, duration of supplementation, and the specific population studied. As with any medical advice or intervention, it's recommended to consult with healthcare professionals for personalized guidance based on individual health conditions. There is a growing recognition of the significance of antioxidant micronutrients such as vitamin C (Vissers et al., 2001; Carcamo et al., 2002) and vitamin E (Ricciarelli et al., 2001) in serving as biological response modifiers. Beyond their traditional role in redox potential, these micronutrients are now acknowledged for their ability to influence functions that extend beyond oxidative processes, encompassing areas such as immunomodulation and gene expression pathways. Additionally, certain compounds labeled as "antioxidants," including polyphenolics, are implicated in similar multifaceted biological responses. Beyond its general immune support, vitamin C exhibits antihistamine effects and can alleviate symptoms associated with influenza, such as runny nose, congestion, sneezing, and inflamed sinuses (Hemila, 1997a,b; Field et al., 2002). The primary role of vitamin C in the immune response against infections lies in its potent antioxidant properties. Ascorbic acid, the active form of vitamin C, serves as a cofactor for various enzymes involved in critical biosynthesis and gene regulation processes (Wintergerst et al., 2006).

Vitamin C mediates the immune response through a multitude of cellular functions within both the acquired and innate immune systems. One notable function is its contribution to the formation of an epithelial barrier, providing protection against various pathogenic organisms. Furthermore, vitamin C enhances the oxidant-scavenging ability of the skin, aiding in the defense against oxidative stress. It also boosts the chemotactic ability of phagocytic cells, thereby increasing the phagocytosis of invading microbes. Importantly, vitamin C plays a pivotal role in the removal of old neutrophils from infection sites, reducing potential damage to infected tissues (Carr and Maggini, 2017).

Vitamin C, a water-soluble vitamin naturally found in various foods, especially citrus fruits, acts as a powerful antioxidant and free radical scavenger, contributing to the enhancement of immune functions. Several clinical studies have demonstrated that an increased intake of vitamin C enhances resistance against various viral and bacterial infections (Dobrange et al., 2019). Vitamin C has proven benefits in the context of upper respiratory infections and severe pneumonia (Gasmi et al., 2020), highlighting that the respiratory syncytial virus, a common cause of both lower and upper respiratory infections, induces the formation of reactive oxygen species (ROS) in the air epithelial cells of the lungs. The resulting ROS leads to pulmonary toxicity by inhibiting lung antioxidants. It has been proposed that administering vitamin C can reduce viral infections by mitigating the harmful effects of ROS. Thus, we can say that vitamin C plays a crucial role in mediating the immune response by influencing various cellular functions within both the acquired and innate immune systems. Its multifaceted involvement underscores its importance in supporting and regulating immune processes.

ROLE OF VITAMIN B COMPLEX IN IMMUNE FUNCTIONS

Vitamin B plays a crucial role in regulating the body's inflammatory response (Morris et al., 2010). Serving as essential cofactors in cellular reactions, B vitamins facilitate the synthesis of amino acids, the fundamental building blocks of antibodies and cytokines. These vitamins are integral to the proliferation and maturation of lymphocytes, key components of the primary immune response (Maggini et al., 2018). Comprising B2, B3, B6, and B12, these water-soluble vitamins function primarily as coenzymes, participating in vital processes within the body. Each B vitamin has a distinct role, contributing significantly to immunity and the ability to counter infections. For instance, vitamin B2, or riboflavin, is specifically involved in cellular energy-yielding metabolic processes (Spinas et al., 2015). Research indicates that exposure to UV light, coupled with vitamin B2, effectively reduces the level of MERS-CoV in the human body (Bashandy et al., 2018).

Vitamin B6 assumes a multifaceted role in immune system function. Notably, it plays a crucial role in activating NK cells, contributing to the body's defense against infections. Additionally, vitamin B6 is involved in the intricate regulation of inflammation, influencing the production of inflammatory cytokines. The adaptive immune system benefits extensively from vitamin B6, engaging in diverse functions ranging from the endogenous synthesis and metabolism of amino acids, the essential building blocks of cytokines and antibodies, to supporting processes like lymphocyte proliferation, differentiation, and maturation. Furthermore, vitamin B6 actively participates in maintaining immune responses by interacting with Th1 cells, a subset of T-helper cells crucial for cell-mediated immunity. Moreover, the vitamin plays a pivotal role in the production of antibodies, essential components of the body's defense mechanism against pathogens. In essence, vitamin B6 emerges as a vital contributor to various facets of the adaptive immune system, ensuring its proper functioning and response to challenges.

Vitamin B12 plays a significant role in the regulation of NK cell functions, contributing to the body's immune defense mechanisms. Beyond its immune-related functions, vitamin B12 is crucial for the proper functioning of various systems, particularly the nervous system. In the realm of cellular and adaptive immunity, vitamin B12 takes on several essential roles. It actively participates in the production of T lymphocytes, a critical component of the immune system responsible for coordinating and regulating immune responses. Moreover, vitamin B12 influences CD8+ T cells, which serve as immunomodulators, contributing to the modulation of immune reactions. The vitamin's impact extends to the intricate realm of single-carbon metabolism, where its interactions with folate are particularly noteworthy. These interactions play a vital role in numerous cellular processes, emphasizing the interconnectedness of vitamins in supporting a robust immune response and overall health (Meydani et al., 1991; Haryanto et al., 2015; Saeed et al., 2016).

Nicotinamide, commonly known as vitamin B3, elevates the eradication of *Staphylococcus aureus* by activating specific genes (Kyme et al., 2012). Computational studies propose that vitamin B12 exerts an inhibitory influence on the RNA-dependent RNA polymerase activity of the SARS-CoV-2 virus, a pivotal enzyme in viral replication (Narayan and Nair, 2020; Wu et al., 2020). Pyridoxal 5′-phosphate, the active form of vitamin B6, plays a multifaceted role in protein, carbohydrate, and lipid metabolism, participating in over a hundred reactions in the body. Recent investigations unveil that bananin (BAN), derived from vitamin B6, exhibits inhibitory effects on the SARS-helicase enzyme, impeding the viral replication process (Tanner et al., 2005). Vitamins B6, B12, and B9 (folic acid) enhance the activity of NK cells, a critical component of antiviral defense (Yoshii et al., 2019). Cumulatively, these findings suggest that B vitamins hold the potential to mitigate complications associated with COVID-19 infection (Tan et al., 2020). In light of the mentioned studies, it becomes evident that vitamin B plays a pivotal role in regulating the inflammatory response, amino acid synthesis, and the proliferation of lymphocytes. Figure 3.1 summarizes the general effects of vitamins on immune modulation during SARS and related infections in conjugation with other components.

CLINICAL STUDIES PERTAINING TO THE ROLE OF VITAMINS IN SARS AND RELATED INFECTIONS

Vitamin C has been extensively studied for its potential benefits in asthma and its association with asthma prevention. In vitro data from endothelial cell lines revealed that vitamin C could inhibit NF-κB activation induced by IL-1 and TNF-α, and it could block the production of IL-8 through mechanisms unrelated to the antioxidant activity of vitamin (Bowie and O'Neill, 2000). In vivo studies employing allergic mouse models of asthma have demonstrated anti-inflammatory and anti-asthmatic effects of vitamin C supplementation. For instance, Jeong and group reported reduced airway hyperresponsiveness (AHR) to methacholine and decreased inflammatory cell infiltration in perivascular and peribronchiolar spaces when vitamin C was supplemented during allergen challenge (Jeong et al., 2010). Another study by Chang et al. found that high-dose vitamin C supplementation in allergen-challenged mice decreased eosinophils in bronchoalveolar lavage fluid (BALF) and shifted the inflammatory pattern to Th1 dominance by increasing the Th1/Th2 cytokine production ratio (Chang et al., 2009).

Experimental and observational studies suggest that vitamin C may play a role in COPD pathogenesis and management. Koike et al. reported that vitamin C supplementation in mice unable to synthesize vitamin C prevented smoke-induced emphysema, restored damaged lung tissue, and decreased oxidative stress caused by smoke-induced emphysema (Koike et al., 2014). A case–control study in Taiwan found that subjects with COPD had lower dietary intake and serum levels of vitamin C compared to healthy controls (Lin et al., 2010).

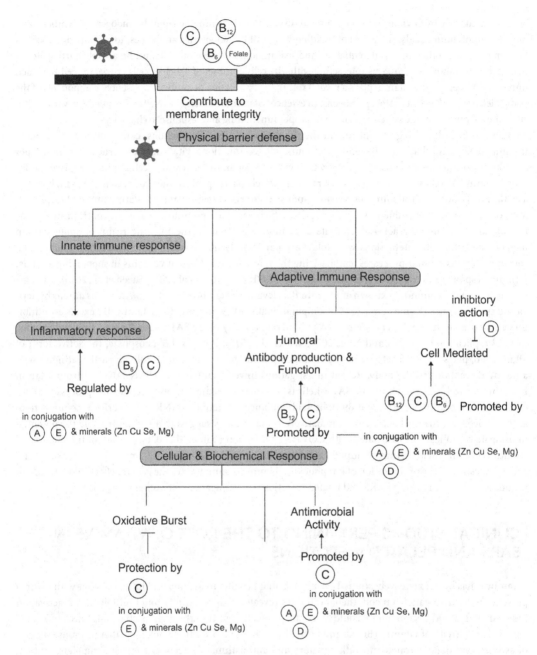

Figure 3.1 Immunomodulatory role of water-soluble vitamins in SARS and related infections. Roles of accessory vitamins (in some cases, fat soluble) or minerals have also been added for reference.

An epidemiological study in the United Kingdom involving over 7,000 adults aged 45–74 years revealed that increased plasma vitamin C concentration was associated with a decreased risk of obstructive airway disease, suggesting a protective effect (Sargeant et al., 2010). In summary, while observational data suggests that vitamin C is crucial for lung health, intervention trials demonstrating efficacy are lacking, and it appears that supplementation with vitamin C-rich whole foods, such as fruits and vegetables, may be more effective.

Human trials have even indicated a significant reduction in pneumonia incidence when individuals were given increased doses of vitamin C in their diet, suggesting its potential to decrease susceptibility to lower respiratory tract infections (Hemilä, 2003). In the context of severe complications, a clinical trial in the USA reported that intravenous (IV) doses of vitamin C reduced the death rate from sepsis-induced acute respiratory distress syndrome (ARDS). Given that the development of ARDS is a critical complication in patients with COVID-19 leading to mortality (Fowler et al., 2019), this finding underscores the potential therapeutic benefits of vitamin C in mitigating severe outcomes.

Recent studies have strongly recommended the consumption of vitamin C as a means to control lower respiratory tract infections, with vitamin C supplementation emerging as a compelling therapeutic intervention for COVID-19 (Colunga Biancatelli et al., 2020a; Liu et al., 2020; Matthay et al., 2020; Zhang et al., 2020a,b). Vitamin C supplementation has shown promising effects in various aspects of respiratory health. In hospitalized elderly patients with acute respiratory infections, administering 200 mg/day of vitamin C led to improvements in respiratory symptoms and an impressive 80% reduction in mortality within the vitamin C group (Hunt et al., 1994). Moreover, several controlled studies have demonstrated a lower incidence of pneumonia in groups supplemented with vitamin C. Furthermore, a meta-analysis of 12 controlled studies involving 1,766 intensive care unit (ICU) patients revealed that oral vitamin C consumption (2.0 g/day) significantly reduced ICU stay duration and mechanical ventilation time. These findings, reported by Hemila et al., suggest potential benefits of vitamin C in ICU settings (Hemila et al., 2013). Also, in avian coronavirus infections, vitamin C has been shown to enhance the resistance of cultured chick embryos (Atherton et al., 1978).

Highlighting the potential therapeutic benefits, a randomized clinical trial (NCT04264533) is currently underway in China. In this trial, approximately 140 individuals infected with SARS-CoV-2 will receive intravenous (IV) vitamin C at a dose of 24 g/day for 7 days (Carr et al., 2017; Carr, 2020). In a notable clinical study by Gorton and Jarvis, supplementation of ascorbic acid resulted in an impressive 85% reduction in flu and cold symptoms among the test groups (Gorton and Jarvis, 1999). Furthermore, a high dose of vitamin C, specifically 12 g/day, significantly improved the conditions of patients suffering from severe acute respiratory tract infections, as reported by Kakodkar et al. (2020). Echoing these findings, yet another study suggested that mega doses of vitamin C can effectively be employed to treat common cold and flu symptoms in children (Banerjee and Kaul, 2010). Beyond specific viral infections, it has also been documented that vitamin C enhances the responsiveness of cells in the immune system, potentially reducing the severity of colds and respiratory infections (Milne, 2008). These findings underscore the potential therapeutic benefits of vitamin C in mitigating symptoms and improving outcomes in various respiratory conditions.

Furthermore, vitamin C plays a crucial role in enhancing the production of interferon α/β, which is a key factor in antiviral immunity during infections (Kim et al., 2011). Another significant mechanism through which vitamin C affects antiviral infections is its free radical scavenging activity. In the investigation conducted by Brinkevich et al. (2012), the 2-O-glycosylated derivatives of ascorbic acid (AA) demonstrated important antiviral properties against herpes simplex virus type 1. This underscores the multifaceted impact of vitamin C on the immune response and its potential as a therapeutic agent against viral infections. Additionally, a derivative of AA, specifically 4,5-unsaturated 4-butyl-substituted 2,3-dibenzyl-L ascorbic acid, demonstrated modest antiviral activity against herpes simplex virus type 2 and coronaviruses, as highlighted in the study by Macan et al. (2019). Notably, the antioxidant activity of vitamin C has been found to inhibit the replication of HIV in chronically infected T cells (Garland and Fawzi, 1999). Further supporting the potential benefits of vitamin C, Colunga Biancatelli et al. (2020b) documented that vitamin C supplementation for USSR soldiers led to a reduction in influenza-related pneumonia viral infections. Clinical trials involving humans have demonstrated a noteworthy decrease in the incidence of pneumonia when higher doses of vitamin C were administered through the diet. These findings indicate that vitamin C holds substantial promise in reducing susceptibility to lower respiratory tract infections (Hemilä, 2003).

FUTURE INSIGHT INTO THE ROLE OF WATER-SOLUBLE VITAMINS IN SARS

Despite controversial studies, there is evidence supporting the idea that supplements rich in water-soluble vitamins may be beneficial in reducing the side effects of diseases. Current scenario emphasizes the need for more investigation and clinical evidence to understand the role of nutrients in defending against infections, especially viral infections. Current knowledge about how nutrients participate in the defense against infections, particularly viral infections, is considered inadequate. There is a recognized need for studying potential synergistic interactions among nutrient supplements and medications, such as antiviral drugs. This suggests that combining nutrient supplements with pharmaceutical interventions may have a positive impact on treatment and recovery. To better understand the relationship between nutrients and viral infections, the investigation of potential synergies between nutrient supplements and antiviral drugs is necessary, as well as gaining insights into the mechanisms of virus transmission and the preventive role of nutrients. This holistic approach could contribute to more effective strategies for the treatment of and recovery from viral infections.

CONCLUSION

The requirements for water-soluble vitamins at all stages of the life cycle need to be determined as they may relate to the prevention of infectious viral diseases like SARS. There are several types of epidemiologic evidence that supports the positive relationship between a diet rich in water-soluble vitamins and enhanced overall health. However, the challenges associated with identifying and delivering sufficient quantities of these vitamins solely through diet, which requires careful supplementation, particularly with vitamin B complex, are yet to be resolved.

REFERENCES

Aleksandrova, K.V., Krisanova, N.V., Ivanchenko, D.G. and Rudko, N.P., 2016. *Biochemistry of Vitamins.* Ukarine: Zaporizhzhya State Medical University.

Atherton, J.G., Kratzing, C.C. and Fisher, A., 1978. The effect of ascorbic acid on infection of chick-embryo ciliated tracheal organ cultures by coronavirus. *Archives of Virology*, 56, pp. 195–199.

Ball, G.F.M. 2013. *Bioavailability and Analysis of Vitamins in Foods.* New York: Springer. https://link.springer.com/book/10.1007/978-1-4899-3414-7#bibliographic-information

Banerjee, D. and Kaul, D., 2010. Combined inhalational and oral supplementation of ascorbic acid may prevent influenza pandemic emergency: a hypothesis. *Nutrition*, 26(1), pp. 128–132.

Barba, F.J., Mariutti, L.R., Bragagnolo, N., Mercadante, A.Z., Barbosa-Cánovas, G.V. and Orlien, V., 2017. Bioaccessibility of bioactive compounds from fruits and vegetables after thermal and nonthermal processing. *Trends in Food Science & Technology*, 67, pp. 195–206.

Bashandy, S.A., Ebaid, H., Abdelmottaleb Moussa, S.A., Alhazza, I.M., Hassan, I., Alaamer, A. and Al Tamimi, J., 2018. Potential effects of the combination of nicotinamide, vitamin B2 and vitamin C on oxidative-mediated hepatotoxicity induced by thioacetamide. *Lipids in Health and Disease*, 17(1), pp. 1–9.

Bates, C.J. and Heseker, H., 1994. Human bioavailability of vitamins: members of EC flair concerted action no. 10: 'measurement of micronutrient apsorption and status'. *Nutrition Research Reviews*, 7(1), pp. 93–127.

Bowie, A.G. and O'Neill, L.A., 2000. Vitamin C inhibits NF-κB activation by TNF via the activation of p38 mitogen-activated protein kinase. *The Journal of Immunology*, 165(12), pp. 7180–7188.

Brinkevich, S.D., Boreko, E.I., Savinova, O.V., Pavlova, N.I. and Shadyro, O.I., 2012. Radical-regulating and antiviral properties of ascorbic acid and its derivatives. *Bioorganic & Medicinal Chemistry Letters*, 22(7), pp. 2424–2427.

Butland, B., Fehily, A. and Elwood, P., 2000. Diet, lung function, and lung function decline in a cohort of 2512 middle aged men. *Thorax*, 55(2), p. 102.

Calderaro, A., Buttrini, M., Farina, B., Montecchini, S., De Conto, F. and Chezzi, C., 2022. Respiratory tract infections and laboratory diagnostic methods: a review with a focus on syndromic panel-based assays. *Microorganisms*, 10(9), p. 1856.

Carbonell-Capella, J.M., Buniowska, M., Barba, F.J., Esteve, M.J. and Frígola, A., 2014. Analytical methods for determining bioavailability and bioaccessibility of bioactive compounds from fruits and vegetables: a review. *Comprehensive Reviews in Food Science and Food Safety*, 13(2), pp. 155–171.

Cárcamo, J.M., Pedraza, A., Bórquez-Ojeda, O. and Golde, D.W., 2002. Vitamin C suppresses TNFα-induced NFκB activation by inhibiting IκBα phosphorylation. *Biochemistry*, 41(43), pp. 12995–13002.

Carr, A.C., 2020. A new clinical trial to test high-dose vitamin C in patients with COVID-19. *Critical Care*, 24(1), pp. 1–2.

Carr, A.C. and Lykkesfeldt, J. 2021. Discrepancies in global vitamin C recommendations: A review of RDA criteria and underlying health perspectives. *Critical Reviews in Food Science and Nutrition*, 61(5), pp. 742–755.

Carr, A.C. and Maggini, S., 2017. Vitamin C and immune function. *Nutrients*, 9(11), p. 1211.

Chang, H.H., Chen, C.S. and Lin, J.Y., 2009. High dose vitamin C supplementation increases the Th1/Th2 cytokine secretion ratio, but decreases eosinophilic infiltration in bronchoalveolar lavage fluid of ovalbumin-sensitized and challenged mice. *Journal of Agricultural and Food Chemistry*, 57(21), pp. 10471–10476.

Chen, B., Tian, E.K., He, B., Tian, L., Han, R., Wang, S., Xiang, Q., Zhang, S., El Arnaout, T. and Cheng, W., 2020. Overview of lethal human coronaviruses. *Signal Transduction and Targeted Therapy*, 5(1), p. 89.

Colunga Biancatelli, R.M.L., Berrill, M. and Marik, P.E., 2020a. The antiviral properties of vitamin C. *Expert Review of Anti-Infective Therapy*, 18(2), pp. 99–101.

Colunga Biancatelli, R.M.L., Berrill, M., Catravas, J.D. and Marik, P.E., 2020b. Quercetin and vitamin C: an experimental, synergistic therapy for the prevention and treatment of SARS-CoV-2 related disease (COVID-19). *Frontiers in Immunology*, 11, p.1451.

Cui, J., Li, F. and Shi, Z.L., 2019. Origin and evolution of pathogenic coronaviruses. *Nature Reviews Microbiology*, 17(3), pp. 181–192. https://doi.org/10.1038/s41579-018-0118-9.

Dobrange, E., Peshev, D., Loedolff, B. and Van den Ende, W., 2019. Fructans as immunomodulatory and antiviral agents: the case of Echinacea. *Biomolecules*, 9(10), p. 615.

Field, C.J., Johnson, I.R. and Schley, P.D., 2002. Nutrients and their role in host resistance to infection. *Journal of Leukocyte Biology*, 71(1), pp. 16–32.

Forastiere, F., Pistelli, R., Sestini, P., Fortes, C., Renzoni, E., Rusconi, F., Dell'Orco, V., Ciccone, G., Bisanti, L. and The SIDRIA Collaborative Group I., 2000. Consumption of fresh fruit rich in vitamin C and wheezing symptoms in children. *Thorax*, 55, pp. 283–288.

Fowler, J.H., Hill, S.J. and Levin, R. 2021. Obradovich N. Stay-at-home orders associate with subsequent decreases in COVID-19 cases and fatalities in the United States. *PLOS ONE*, 16(6), e0248849.

Funk, C., 1912. The etiology of the deficiency diseases. Beri-beri, polyneuritis in birds, epidemic dropsy, scurvy, experimental scurvy in animals, infantile scurvy, ship beri-beri, pellagra. *Journal of State Medicine*, 20, pp. 341– 368.

Garland, M. and Fawzi, W.W., 1999. Antioxidants and progression of human immunodeficiency virus (HIV) disease. *Nutrition Research*, 19(8), pp. 1259–1276.

Gasmi, A., Noor, S., Tippairote, T., Dadar, M., Menzel, A. and Bjørklund, G., 2020. Individual risk management strategy and potential therapeutic options for the COVID-19 pandemic. *Clinical Immunology*, 215, p. 108409.

Gorton, H.C. and Jarvis, K., 1999. The effectiveness of vitamin C in preventing and relieving the symptoms of virus-induced respiratory infections. *Journal of Manipulative and Physiological Therapeutics*, 22(8), pp. 530–533.

Gregory, J.F. 3rd, 1998. Nutritional properties and significance of vitamin glycosides. *Annual Reviews of Nutrition*, 18, pp. 277–296.

Haryanto, B., Suksmasari, T., Wintergerst, E. and Maggini, S., 2015. Multivitamin supplementation supports immune function and ameliorates conditions triggered by reduced air quality. *Vitamins & Minerals*, 4(2), pp. 2376–1318.

Hemilä, H., 1997a. Vitamin C intake and susceptibility to pneumonia. *The Pediatric Infectious Disease Journal*, 16(9), pp. 836–837.

Hemilä, H., 1997b. Vitamin C intake and susceptibility to the common cold. *British Journal of Nutrition*, 77(1), pp. 59–72.

Hemilä, H., 2003. Vitamin C and SARS coronavirus. *Journal of Antimicrobial Chemotherapy*, 52(6), pp. 1049–1050.

Hemilä, H. and Chalker E. 2013. Vitamin C for preventing and treating the common cold. *Cochrane Database of Systematic Reviews*, 2013(1), CD000980. doi: 10.1002/14651858.CD000980.pub4.

Hu, B., Guo, H., Zhou, P. and Shi, Z.L., 2021. Characteristics of SARS-CoV-2 and COVID-19. *Nature Reviews Microbiology*, 19(3), pp. 141–154. https://doi.org/10.1038/s41579-020-00459-7.

Huang, C.L., Wang, Y.M., Li, X.W., Ren, L.L., Zhao, J.P., Hu, Y., Zhang, L., Fan, G.H., Xu, J.Y., Gu, X.Y., et al., 2020. Clinical features of patients infected with 2019 novel coronavirus in Wuhan, China. *Lancet*, 395(10223), pp. 497–506. https://doi.org/10.1016/S0140-6736(20)30183-5.

Hunt, C., Chakravorty, N.K., Annan, G., Habibzadeh, N. and Schorah, C.J., 1994. The clinical effects of vitamin C supplementation in elderly hospitalised patients with acute respiratory infections. *International Journal for Vitamin and Nutrition Research*, 64(3), pp. 212–219.

Jeong, Y.J., Kim, J.H., Kang, J.S., Lee, W.J. and Hwang, Y.I., 2010. Mega-dose vitamin C attenuated lung inflammation in mouse asthma model. *Anatomy & Cell Biology*, 43(4), pp. 294–302.

Kakodkar, P., Kaka, N. and Baig, M.N., 2020. A comprehensive literature review on the clinical presentation, and management of the pandemic coronavirus disease 2019 (COVID-19). *Cureus*, 12(4), e7560.

Kelly, F.J., 2005. Vitamins and respiratory disease: antioxidant micronutrients in pulmonary health and disease. *Proceedings of the Nutrition Society*, 64(4), pp. 510–526.

Kim, J.K., Cho, M.L., Karnjanapratum, S., Shin, I.S. and You, S.G., 2011. In vitro and in vivo immunomodulatory activity of sulfated polysaccharides from *Enteromorpha prolifera*. *International Journal of Biological Macromolecules*, 49(5), pp. 1051–1058.

Koike, K., Ishigami, A., Sato, Y., Hirai, T., Yuan, Y., Kobayashi, E., Tobino, K., Sato, T., Sekiya, M., Takahashi, K. and Fukuchi, Y., 2014. Vitamin C prevents cigarette smoke-induced pulmonary emphysema in mice and provides pulmonary restoration. *American Journal of Respiratory Cell and Molecular Biology*, 50(2), pp. 347–357.

Kyme, P., Thoennissen, N.H., Tseng, C.W., Thoennissen, G.B., Wolf, A.J., Shimada, K., Krug, U.O., Lee, K., Muller-Tidow, C., Berdel, W.E. and Hardy, W.D., 2012. C/EBPε mediates nicotinamide-enhanced clearance of Staphylococcus aureus in mice. *The Journal of Clinical Investigation*, 122(9), pp. 3316–3329.

Lin, Y.C., Wu, T.C., Chen, P.Y., Hsieh, L.Y. and Yeh, S.L., 2010. Comparison of plasma and intake levels of antioxidant nutrients in patients with chronic obstructive pulmonary disease and healthy people in Taiwan: a case-control study. *Asia Pacific Journal of Clinical Nutrition*, 19(3), pp. 393–401.

Liu, F., Zhu, Y., Zhang, J., Li, Y. and Peng, Z., 2020. Intravenous high-dose vitamin C for the treatment of severe COVID-19: study protocol for a multicentre randomised controlled trial. *BMJ Open* 2020, 10, p. e039519.

Macan, A.M., Harej, A., Cazin, I., Klobučar, M., Stepanić, V., Pavelić, K., Pavelić, S.K., Schols, D., Snoeck, R., Andrei, G. and Raić-Malić, S., 2019. Antitumor and antiviral activities of 4-substituted 1, 2, 3-triazolyl-2, 3-dibenzyl-L-ascorbic acid derivatives. *European Journal of Medicinal Chemistry*, 184, p. 111739.

Maggini, S., Pierre, A., and Calder, P.C., 2018. Immune function and micronutrient requirements change over the life course. *Nutrients*, 10, p. 1531.

Matthay, M.A., Aldrich, J.M. and Gotts, J.E., 2020. Treatment for severe acute respiratory distress syndrome from COVID-19. *The Lancet Respiratory Medicine*, 8(5), pp. 433–434.

Meydani, S.N., Ribaya-Mercado, J.D., Russell, R.M., Sahyoun, N., Morrow, F.D. and Gershoff, S.N., 1991. Vitamin B6 deficiency impairs interleukin 2 production and lymphocyte proliferation in elderly adults. *The American Journal of Clinical Nutrition*, 53(5), pp. 1275–1280.

Milne, A., 2008. Summary of 'Vitamin C for preventing and treating the common cold'. *Evidence-Based Child Health: A Cochrane Review Journal*, 3(3), pp. 721–722.

Morris, M.S., Sakakeeny, L., Jacques, P.F., Picciano, M.F. and Selhub, J., 2010. Vitamin B-6 intake is inversely related to, and the requirement is affected by, inflammation status. *Journal of Nutrition*, 140, pp. 103–110.

Narayanan, N. and Nair, D.T. 2020. Vitamin B12 may inhibit RNA-dependent-RNA polymerase activity of nsp12 from the SARS-CoV-2 virus. *IUBMB Life*. 72(10), pp. 2112–2120.

Peters, R.A., 1963. *Biochemical Lesions and Lethal Synthesis*. (No Title). Oxford: Pergamon Press.

Rahman, I., 2003. Oxidative stress, chromatin remodeling and gene transcription in inflammation and chronic lung diseases. *BMB Reports*, 36(1), pp. 95–109.

Ricciarelli, R., Zingg, J.M. and Azzi, A., 2001. Vitamin E: protective role of a Janus molecule. *The FASEB Journal*, 15(13), pp. 2314–2325.

Rota, P.A., M.S. Oberste, S.S. Monroe, W.A. Nix, R. Campagnoli, J.P. Icenogle, S. Penaranda, B. Bankamp, K. Maher, M.H. Chen, et al. 2003. Characterization of a novel coronavirus associated with severe acute respiratory syndrome. *Science*, 300(5624), pp. 1394–1399. https://doi.org/10.1126/science.1085952.

Saeed, F., Nadeem, M., Ahmed, R.S., Tahir Nadeem, M., Arshad, M.S. and Ullah, A., 2016. Studying the impact of nutritional immunology underlying the modulation of immune responses by nutritional compounds-a review. *Food and Agricultural Immunology*, 27(2), pp. 205–229.

Sargeant, L., Jaeckel, A. and Wareham, N. 2000. Interaction of vitamin C with the relation between smoking and obstructive airways disease in EPIC Norfolk. *The European Respiratory Journal*, 16, pp. 397–402.

Schwartz, J. and Weiss, S.T., 1994. Relationship between dietary vitamin C intake and pulmonary function in the First National Health and Nutrition Examination Survey (NHANES I). *The American Journal of Clinical Nutrition*, 59(1), pp. 110–114.

Shaheen, S.O., Sterne, J.A., Thompson, R.L., Songhurst, C.E., Margetts, B.M., and Burney, P.G., 2001. Dietary antioxidants and asthma in adults: Population-based case-control study. *American Journal of Respiratory and Critical Care Medicine*, 164, pp. 1823–1828.

Spinas, E., Saggini, A., Kritas, S.K., Cerulli, G., Caraffa, A., Antinolfi, P., Pantalone, A., Frydas, A., Tei, M., Speziali, A. and Saggini, R., 2015. Crosstalk between vitamin B and immunity. *Journal of Biological Regulators and Homeostatic Agents*, 29(2), pp. 283–288.

Su, S., et al., 2016. Epidemiology, genetic recombination, and pathogenesis of coronaviruses. *Trends in Microbiology*, 24, pp. 490–502.

Tan, C.W., Ho, L.P., Kalimuddin, S., Cherng, B.P.Z., Teh, Y.E., Thien, S.Y., Wong, H.M., Tern, P.J.W., Chandran, M., Chay, J.W.M. and Nagarajan, C., 2020. A cohort study to evaluate the effect of combination Vitamin D, Magnesium and Vitamin B12 (DMB) on progression to severe outcome in older COVID-19 patients. *Nutrition*, pp. 79–80, 111017.

Tanner, J.A., Zheng, B.J., Zhou, J., Watt, R.M., Jiang, J.Q., Wong, K.L., Lin, Y.P., Lu, L.Y., He, M.L., Kung, H.F. and Kesel, A.J., 2005. The adamantane-derived bananins are potent inhibitors of the helicase activities and replication of SARS coronavirus. *Chemistry & Biology*, 12(3), pp. 303–311.

Varraso, R., 2012. Nutrition and asthma. *Current Allergy and Asthma Reports*, 12, pp. 201–210.

van der Hoek, L. 2007. Human coronaviruses: what do they cause? *Antiviral Therapy*, 12(4 Pt B), pp. 651–658.

Vissers, M.C., Lee, W.G. and Hampton, M.B., 2001. Regulation of apoptosis by vitamin C: specific protection of the apoptotic machinery against exposure to chlorinated oxidants. *Journal of Biological Chemistry*, 276(50), pp. 46835–46840.

Wang, X., Li, Y., Shi, T., Bont, L.J., Chu, H.Y., Zar, H.J., Wahi-Singh, B., Ma, Y., Cong, B., Sharland, E., Riley, R.D., Deng, J., Figueras-Aloy, J., Heikkinen, T., Jones, M.H., Liese, J.G., Markić, J., Mejias, A., Nunes, M.C., Resch, B., Satav, A., Yeo, K.T., Simões, E.A.F. and Nair, H., 2024. Respiratory virus global epidemiology network; RESCEU investigators. Global disease burden of and risk factors for acute lower respiratory infections caused by respiratory syncytial virus in preterm infants and young children in 2019: a systematic review and meta-analysis of aggregated and individual participant data. *Lancet*, 403(10433), pp. 1241–1253.

WHO, 2019. Middle East respiratory syndrome coronavirus (MERS-CoV). *MERS Monthly Summary, November*. https://www.who.int/emergencies/disease-outbreak-news/item/2024-DON516#:~:text=Overall%2C%20a%20total%20of%202613,newly%20reported%20cases%20and%20death

Wintergerst, E.S., Maggini, S. and Hornig, D.H., 2006. Immune-enhancing role of vitamin C and zinc and effect on clinical conditions. *Annals of Nutrition and Metabolism*, 50(2), pp. 85–94.

Wolfel, R., Corman, V.M., Guggemos, W., Seilmaier, M., Zange, S., Muller, M.A., Niemeyer, D., Jones, T.C., Vollmar, P., Rothe, C., et al., 2020. Virological assessment of hospitalized patients with COVID_2019. *Nature*, 581(7809), pp. 465–469. https://doi.org/10.1038/s41586-020-2196-x

Wood, L.G., Gibson, P.G. and Garg, M.L., 2003. Biomarkers of lipid peroxidation, airway inflammation and asthma. *European Respiratory Journal*, 21(1), pp. 177–186.

Wu, C., Liu, Y., Yang, Y., Zhang, P., Zhong, W., Wang, Y., Wang, Q., Xu, Y., Li, M., Li, X. and Zheng, M., 2020. Analysis of therapeutic targets for SARS-CoV-2 and discovery of potential drugs by computational methods. *Acta Pharmaceutica Sinica* B, 10(5), pp. 766–788.

Yoshii, K., Hosomi, K., Sawane, K. and Kunisawa, J., 2019. Metabolism of dietary and microbial vitamin B family in the regulation of host immunity. *Frontiers in Nutrition*, 6, p. 48.

Zaki, A.M., van Boheemen, S., Bestebroer, T.M., Osterhaus, A.D. and Fouchier, R.A. 2012. Isolation of a novel coronavirus from a man with pneumonia in Saudi Arabia. *The New England Journal of Medicine*, 367(19), pp. 1814–1820.

Zelka, F.Z., Kocatürk, R.R., Özcan, Ö.Ö. and Karahan, M., 2022. Can nutritional supports beneficial in other viral diseases be favorable for COVID-19? *Korean Journal of Family Medicine*, 43(1), p. 3.

Zhang, J., Xie, B. and Hashimoto, K., 2020b. Current status of potential therapeutic candidates for the COVID-19 crisis. *Brain, Behavior, and Immunity*, 87, pp. 59–73.

Zhang, L., Lin, D., Sun, X., Curth, U., Drosten, C., Sauerhering, L., Becker, S., Rox, K. and Hilgenfeld, R., 2020a. Crystal structure of SARS-CoV-2 main protease provides a basis for design of improved α-ketoamide inhibitors. *Science*, 368(6489), pp. 409–412.

Zheng, J. 2020. SARS-CoV-2: An emerging Coronavirus that causes a global threat. *International Journal of Biological Sciences*, 16(10), pp. 1678–1685.

Zhou, P., Yang, X.L., Wang, X.G., Hu, B., Zhang, L., Zhang, W., Si, H.R., Zhu, Y., Li, B., Huang, C.L., Chen, H.D., Chen, J., Luo, Y., Guo, H., Jiang, R.D., Liu, M.Q., Chen, Y., Shen, X.R., Wang, X., Zheng, X.S., Zhao, K., Chen, Q.J., Deng, F., Liu, L.L., Yan, B., Zhan, F.X., Wang, Y.Y., Xiao, G.F. and Shi, Z.L., 2020. A pneumonia outbreak associated with a new coronavirus of probable bat origin. *Nature*, 579(7798), pp. 270–273. https://doi.org/10.1038/s41586-020-2012-7.

4

Role of fat-soluble vitamins in combating Middle East Respiratory Syndrome (MERS)

VIGNESH SOUNDERRAJAN, SAM EBENEZER RAJADAS,
T. THANGAM, SUDHANARAYANI S. RAO, SAKTHIVEL JEYARAJ,
AND KRUPAKAR PARTHASARATHY

INTRODUCTION

Middle East respiratory syndrome (MERS) is a viral respiratory illness caused by the Middle East respiratory syndrome coronavirus (MERS-CoV), first identified in Saudi Arabia in 2012. Belonging to the Coronaviridae family, similar to SARS-CoV and SARS-CoV-2, MERS-CoV is primarily transmitted to humans from dromedary camels through close contact (WHO, 2013). Human-to-human transmission occurs via respiratory droplets or close contact with an infected person. Symptoms range from mild to severe, including fever, cough, shortness of breath, and sometimes gastrointestinal symptoms, with severe cases progressing to pneumonia, Acute Respiratory Distress Syndrome (ARDS), kidney failure, and death. The incubation period is typically 2–14 days (Omrani and Shalhoub, 2015). Those at higher risk of severe illness include individuals with underlying medical conditions and weakened immune systems, as well as older adults (Jahan and Al Maqbali, 2015). While most cases have been in the Arabian Peninsula, cases have been reported globally due to travel-related transmission. Prevention involves hand hygiene, avoiding sick individuals, and camel contact (Ramadan and Shaib, 2019). There is no specific treatment, but supportive care helps manage symptoms. Ongoing research aims to understand transmission and improve surveillance for early detection and containment. Though not as easily spread as some respiratory viruses, MERS-CoV remains a concern due to its potential for severe illness and mortality, particularly in vulnerable populations (Badawi and Ryoo, 2016). The role of vitamins in combating MERS specifically hasn't been extensively studied, but there is general scientific evidence regarding the importance of vitamins in supporting the immune system, which could potentially aid in fighting off respiratory infections, including MERS.

Vitamins are essential micronutrients that must be incorporated into the food as our physiological system cannot synthesize it, except vitamin D. They are the components of normal health that facilitate vital functions like immune defense and metabolism. Vitamins have a well-established role in treating several diseases. Due to their solubility in organic solvents and resemblance to fats in terms of absorption and transportation, vitamins A, D, E, and K are known as fat-soluble vitamins. Over the past few decades, plenty of research has demonstrated the critical role of vitamin A in immunity against viral diseases. Its ways of regulation were observed to be better than the micronutrients. However, the mechanism of regulation is unclear so far. Oral vitamin A supplementation or fortification has been shown in community- and hospital-based clinical trials to reduce preschool child mortality in underdeveloped nations by approximately one-third (Chelstowska et al., 2016). Several fat-soluble vitamins, including A, D, E, and K, have demonstrated some potential benefits in the treatment of COVID-19. Incorporating nutritional therapy into patient care is imperative for both better and quicker recovery from this potentially fatal respiratory viral infection. Above all, ensuring that malnutrition is monitored for and providing the best possible nutritional supplements are essential for the immune system to function at its best. The rapid outbreak and spread of the COVID-19 like pandemic and other influenza viruses are exaggerated by their severity in pathogenesis. Though this respiratory virus has spread rapidly, there are currently few evidence-based therapeutic options available. It's important to note that while vitamins play crucial roles in supporting immune function, supplementation should not be seen as

DOI: 10.1201/9781003435686-4

a substitute for other preventive measures, such as vaccination, good hygiene practices, and avoiding exposure to infectious agents. Additionally, individual vitamin requirements can vary, and excessive intake of certain vitamins can have adverse effects. Overall, while vitamins may contribute to overall immune health and potentially aid in combating respiratory infections like MERS, more research specifically focused on their role in MERS prevention and treatment is needed to draw definitive conclusions.

In the meantime, a number of studies have been initiated to identify the function of various fat-soluble vitamins and minerals in preventing respiratory viral infection. This chapter provides insights into the effects of fat-soluble vitamins in modulating immune responses to viral infections.

VITAMIN A AND PATHOPHYSIOLOGICAL ROLES

Vitamin A is an essential fat-soluble vitamin for eyesight and immune regulation. The major active forms of vitamin A are retinol, retinal, and retinoic acid which have potential anti-inflammatory properties. It has been used as an anti-inflammatory agent in the treatment regimen of several infectious diseases as an anti-inflammatory agent (Li et al., 2020). A group of carotenoids, namely β-carotene, also known as pro-vitamin A, are transformed into retinol in the intestine and then absorbed. Retinoic acid is the vitamin A form with the most bioactive structure among the others. The retinoic acid form of vitamin A upregulates the cytokines that subside the inflammation and IgA antibodies during viral infections like measles and flu (Al-Sumiadai et al., 2020).

To further support this finding, a recent study reported that a deficiency in vitamin A led to excessive inflammation and increased susceptibility to viral infections (Timoneda et al., 2018). Furthermore, retinoic acid depletion is a prevalent phenomenon during inflammatory disorder-inducing diseases like COVID-19 (Sarohan, 2020), wherein the immune system collapses due to inhibition of the type 1 interferon pathway (Devasthanam, 2014). Vitamin A supplementation could lower the death rate of children infected with the virus. Treating Human Immunodefeciency Virus (HIV)-positive individuals with antiretroviral medications and vitamin A supplements was also highly successful (Chelstowska et al., 2016). Most studies have found that vitamin A supplements may improve immunity in virus-infected individuals by focusing on T-cell and B-cell functioning. Vitamin A prevents the measles virus from spreading by regulating the immune response of uninfected cells and protecting them from the infection while the virus replicates by causing the production of interferon (IFN). The innate immune response to viral infections is significantly influenced by the interferon cell signaling pathway. The pathways that are crucial for the replication of DNA viruses, like mitogen-activated protein kinase (MAPK) and nuclear factor kappa-light-chain-enhancer of activated B cells (NF-κB) pathways, are interfered by β-carotene (Kumar et al., 2018). Retinol, retinoic acid, and retinol are the three main active forms of vitamin A. Their regular intake of vitamin A is essential to enhance the body's defense mechanisms against infections, and hence, they are considered "anti-infectives." The activity of vitamin A against the infectious bronchitis virus (IBV), a type of coronavirus that was more prevalent in chickens fed a diet low in vitamin A than in those given adequate amounts of the vitamin, was another evidence of the important relationship between vitamin A and COVID-19 infection. Furthermore, retinoids are closely related compounds to vitamin A that possess strong immunomodulating properties and the ability to increase and intensify the effectiveness of IFN-γ (Chelstowska et al., 2016). Vitamin A interacts with the host interleukin 10, MAPK 1, intercellular adhesion molecule 1, epidermal growth factor receptor, catalase, MAPK 14, and protein kinase C β for immune modulation (Li et al., 2020). As a result, a number of studies have suggested that vitamin A and its metabolites may be helpful in the treatment of COVID-19 (Al-Sumiadai et al., 2020). Vitamin A reduces the inflammation and fibrosis caused by SARS-CoV infections (Timoneda et al., 2018).

VITAMIN E AND PATHOPHYSIOLOGICAL ROLES

Vitamin E is a fat-soluble vitamin that includes four tocopherols and four tocotrienols/tocopherols with a potent antioxidant activity and enhances immunity. This tocotrienol compound can work in cellular membranes as a free radical scavenger, supporting regular immunological activity. According to Lee and Han

(2018), the chromanol ring in vitamin E is responsible for inhibiting the oxidation of polyunsaturated fatty acids, rendering it an antioxidant. By absorbing the lipid peroxyl radicals and preventing them from oxidizing the nearby fatty acid chains, vitamin E functions as a chain breaker of polyunsaturated fatty acids (PUFAs) in the lipid membranes (Comporti, 1993). Vitamins C and E, for example, have been shown to serve as antioxidants and anti-inflammatory agents, hence suggesting these vitamins as effective treatments for COVID-19 (Gasmi et al., 2020).

VITAMIN D AND PATHOPHYSIOLOGICAL ROLES

Vitamin D is produced by the conversion of 7-dehydrocholesterol by exposure to sunlight which stimulates the skin to produce a secosteroid hormone. Though fortified drinks, milk, eggs, and fatty fish are good sources of vitamin D, the average diet isn't much enough to compensate. The transformation of the inactive form of vitamin D (7-dehydrocholesterol) into the active form (1,25 dihydroxy vitamin D3) is mediated by the liver. This conversion facilitates the gut's absorption of calcium (Schwalfenberg, 2011), a major regulator of the intracellular signaling system and the transcription of viral genes. Vitamin D influences the expression of antimicrobial peptides like cathelicidin which has strong antiviral properties (Vyas et al., 2020).

IMPACT OF VITAMIN D ON MERS AND RELATED INFECTIONS

Seasonal vitamin D deficiency due to reduced exposure to the sun was found to increase the incidence of seasonal influenza viral infection by affecting the innate immune system, thereby promoting replication of the virus in the infected host. Although many previously intriguing facts regarding the epidemiology of influenza have been explained by the vitamin D insufficiency hypothesis, there are still gaps in our knowledge of the etiology and clinical presentations of the disease (Brockman-Schneider et al., 2020). The activated vitamin D3 modulates the immune system by decreasing the expression of inflammatory cytokines and enhancing macrophage function. Additionally, vitamin D increases the expression of potent antimicrobial peptides (AMPs), which are found in natural killer cells, neutrophils, monocytes, and respiratory tract epithelial cells. Immune modulation in the form of increased chemokine activity and recruitment of macrophages, neutrophils, and T cells near the human rhinovirus RV-16-infected cells was observed when the cells were nourished with vitamin D supplementation (Drysdale et al., 2014). There are several ways that vitamin D works, including interacting with nonspecific defensive systems, activating toll-like receptors, and/or raising cathelicidin and β-defensin levels. This affects acquired immunity by reducing the amount of immunoglobulin that plasma cells release and the generation of pro-inflammatory cytokines, which in turn modulates T-cell activity. Positive outcomes also indicated that vitamin D supplements are crucial for the treatment of respiratory tract infections like MERS.

VITAMIN K AND PATHOPHYSIOLOGICAL ROLES

Vitamin K, one of the fat-soluble vitamins, is needed to synthesize numerous proteins – factor II (prothrombin) and factors VII, IX, and X involved in regulating blood clotting (coagulation) (McNally et al., 2014). Two distinct forms of natural vitamin K can be found in food: K1 (phylloquinone), which is primarily found in green leafy vegetables and is an efficient form of dietary vitamin K, and K2 (menaquinones) (Brody, 1999). Researchers have now noticed a relationship between COVID-19 outcomes and vitamin K levels. One of the main indicators of poor outcomes for patients with sepsis brought on by an infection is coagulopathy. Tang et al. observed coagulopathy in 183 consecutive patients with severe COVID-19, and this condition is linked to a poor prognosis (Tang et al., 20204). Disseminated intravascular coagulation (DIC) and coagulopathy seem to be linked to increased death rates. Additionally, it seems that low vitamin K levels are linked to accelerated elastin breakdown, thus destroying lung tissue, which makes COVID-19 patients have trouble breathing (Crowther et al., 2002). Low vitamin K levels are thought to be connected to COVID-19 severity since patients with severe disease are more likely to have co-morbidities like type II diabetes, hypertension, and cardiovascular disorders, all of which are linked to lower vitamin K levels. Given that patients admitted to the intensive care unit (ICU) have a risk of vitamin K deficiency, the supplements of it given to patients at the time of admission may help lower the risk of vitamin K deficiency and subsequent consequences (Riphagen et al., 2017). Additionally, extended use of broad-spectrum antibiotics, such as cephalosporins, might disrupt intestinal flora's ability to synthesize vitamin K and reduce the amount of vitamin K absorbed by changing gastrointestinal functioning. These factors should be taken into account while treating an infection.

CLINICAL STUDIES EXPLORING VITAMIN D SUPPLEMENTATION IN MERS

High maternal serum vitamin D levels have been linked to protection against respiratory viral infection in both human and animal studies. This suggests that high vitamin D levels during pregnancy may shield the unborn child from Acute Respiratory Infections (ARIs) or virally induced wheezing episodes. High cord blood 25 (OH) D levels have been linked to a lower incidence of respiratory infections and childhood wheeze (Ginde et al., 2009). Research on 743 children (3–15 years old) done between December 2008 and January 2009 found a correlation between viral respiratory tract infection and low serum 25 (OH) vitamin D levels. Thirty percent of the subjects had at least one viral respiratory tract infection, as determined by polymerase chain reaction from the nasopharyngeal samples. The risk of viral respiratory tract infections was found to be 50% higher at serum 25 (OH) D levels below 30 ng/mL (75 nmol/L) and 70% higher at levels below 20 ng/mL (50 nmol/L). Research on 743 children (3–15 years old) done between December 2008 and January 2009 found a correlation between viral respiratory tract infection and low serum 25 (OH) vitamin D levels. Thirty percent of the subjects had at least one viral respiratory tract infection, as determined by polymerase reaction from the nasopharyngeal samples. The risk of viral Respiratory Tract Infections (RTIs) was found to be 50% higher at serum 25 (OH) D levels below 30 ng/mL (75 nmol/L) and 70% higher at levels below 20 ng/mL (50 nmol/L) (Science et al., 2013) (Figure 4.1).

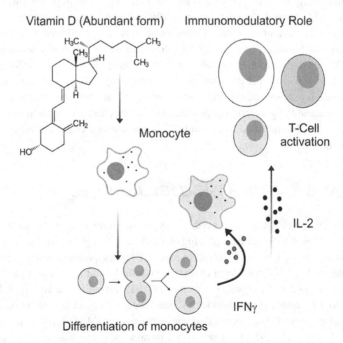

Figure 4.1 Role of vitamins in inducing innate immunity. Vitamin D improves macrophage differentiation, antimicrobial peptides, and chemotaxis to activate the innate immune responses.

ROLE OF LIPID-SOLUBLE VITAMINS IN INFLUENZA

It is possible to consider the severity of an influenza virus infection as a consequence of two sets of opposing forces. The elements that cause infection are on one side and exert inhibitory effects on the virus at different stages of the disease process. The elements that cause the disease, as well as those which cause the virus to proliferate, disseminate, and cause symptoms, are on the opposite side. There is proof that retinoids are essential to both processes. According to certain theories, decreased exposure to sunlight and/or preexisting vitamin D deficiency both increase the build-up and possible toxicity of endogenous retinoids, while a lower ratio of vitamin D to vitamin A causes viral activation or makes people more susceptible to new influenza virus strains; increased but normal physiological concentrations of retinoid successfully inhibit the pathogenesis of influenza in conjunction with vitamin D. Higher background concentrations of vitamin A (i.e., very low vitamin D:A ratios), for example, are linked to liver-related diseases and conditions that promote viral replication and raise the risk of serious or fatal disease complications (Mawson, 2013).

CONCLUSION

In a study, levels of vitamins A, D, and E in children undergoing normal physical examination and children with recurrent respiratory infections were explored (Zhang et al., 2019). Both the active group and the stable group had significantly lower levels of vitamins A, D, and E; then, the control group and the active group levels were much lower than the stable group's. Vitamin A is a fat-soluble vitamin that is vital to human health and plays a major role in immune system maintenance (Niki, 2014). Severe viral disorders are caused by vitamin A deficiency, and chronic vitamin A deficiency directly raises the mortality rate. Vitamin D controls endocrine function and prevents the growth of malignant cells, and maintains homeostasis of hematological system (Aglipay et al., 2017). The clinical outcome for individuals with respiratory infections is significantly influenced by vitamin D supplementation. Supplementing with vitamin E, a potent antioxidant that is equally vital to the body's physiological processes, helps to enhance cellular immunity by stabilizing the structure of cell membranes and preventing lipid peroxidation (Habibian et al., 2014). Individuals with respiratory infections have low amounts of vitamins D, E, and A. A rapid solution for managing viral infections that are currently on increasing incidence is to repurpose vitamins as antiviral medicines. A study found that methylcobalamin and hydroxocobalamin may have broad-spectrum inhibitory effects on the three CoVs that were investigated (Moatasim et al., 2023). These promising vitamins were the subject of in silico studies to examine their interactions with viral proteins (S-RBD, 3CL pro, and RdRp), as well as cell receptors (ACE2, DPP4, and hAPN protein) specific to SARS-CoV-2, MERS-CoV, and HCoV-229E. The results suggested that hydroxocobalamin, methylcobalamin, and cyanocobalamin may have a strong binding affinity to these proteins (Crimi et al., 2004). These findings suggest that people infected with the coronavirus may benefit from methylcobalamin. Patients with infections who supplemented vitamin E are likely to benefit from improved immune function and increased resistance to infection, which will reduce mortality. It has been observed that vitamin C restores vitamin E's antioxidant capacity and that vitamins E and C together, when given to the viral-infected patients, including those with COVID-19 infection, have been shown to be effective. There is no literature found related to the role of fat-soluble vitamins in the context of MERS. The clinical data's related to acute respiratory viral disease would be co-related to the MERS.

REFERENCES

Aglipay M, Birken CS, Parkin PC, Loeb MB, Thorpe K, Chen Y, Laupacis A, Mamdani M, Macarthur C, Hoch JS, Mazzulli T. Effect of high-dose vs standard-dose wintertime vitamin D supplementation on viral upper respiratory tract infections in young healthy children. *JAMA*. 2017;318(3):245–54.

Al-Sumiadai MM, Ghazzay H, Al-Dulaimy WZ. Therapeutic effect of Vitamin A on severe COVID-19 patients. *EurAsian Journal of BioSciences*. 2020;14(2):7347–50.

Badawi A, Ryoo SG. Prevalence of comorbidities in the Middle East respiratory syndrome coronavirus (MERS-CoV): a systematic review and meta-analysis. *International Journal of Infectious Diseases*. 2016;49:129–33.

Brockman-Schneider RA, Pickles RJ, Gern JE. Effects of vitamin D on airway epithelial cell morphology and rhinovirus replication. *PLOS ONE* 2014;9:e86755.

Brody T. *Classification of Biological Structure*. Academic Press: San Diego, 1999, pp. 1–56.

Chelstowska S, Widjaja-Adhi MA, Silvaroli JA, Golczak M. Molecular basis for vitamin A uptake and storage in vertebrates. *Nutrients*. 2016;8(11):676.

Comporti M. Lipid peroxidation. An overview. In: G. Poli, E. Albano, and M. U. Dianzani (eds.), *Free Radicals: From Basic Science to Medicine*, 1993, pp. 65–79. Basel: Birkhäuser.

Crimi E, Liguori A, Condorelli M, Cioffi M, Astuto M, Bontempo P, Pignalosa O, Vietri MT, Molinari AM, Sica V, Della Corte F. The beneficial effects of antioxidant supplementation in enteral feeding in critically ill patients: a prospective, randomized, double-blind, placebo-controlled trial. *Anesthesia & Analgesia*. 2004;99(3):857–63.

Crowther MA, McDonald E, Johnston M, Cook D. Vitamin K deficiency and D-dimer levels in the intensive care unit: a prospective cohort study. *Blood Coagulation & Fibrinolysis*. 2002;13(1):49–52.

Devasthanam AS. Mechanisms underlying the inhibition of interferon signaling by viruses. *Virulence*. 2014;5(2):270–7.

Drysdale SB, Prendergast M, Alcazar M, Wilson T, Smith M, Zuckerman M, Broughton S, Rafferty G, Johnston SL, Hodemaekers HM, et al. Genetic predisposition of RSV infection-related respiratory morbidity in preterm infants. *European Journal of Nuclear Medicine and Molecular Imaging*. 2014;173:905–12.

Gasmi A, Tippairote T, Mujawdiya PK, Peana M, Menzel A, Dadar M, Benahmed AG, Björklund G. Micronutrients as immunomodulatory tools for COVID-19 management. *Clinical Immunology*. 2020;220:108545.

Ginde, A.A., Liu, M.C. and Camargo, C.A. Jr. 2009. Demographic differences and trends of vitamin D insufficiency in the US population, 1988-2004. *Archives of Internal Medicine*, 169(6), pp. 626–632.

Habibian M, Ghazi S, Moeini MM, Abdolmohammadi A. Effects of dietary selenium and vitamin E on immune response and biological blood parameters of broilers reared under thermoneutral or heat stress conditions. *International Journal of Biometeorology*. 2014;58:741–52.

Jahan F, Al Maqbali AA. The Middle East Respiratory Syndrome Coronavirus (MERS-COV). *Middle East Journal of Family Medicine*. 2015;13(1):26–30.

Kumar R, Khandelwal N, Thachamvally R, Tripathi BN, Barua S, Kashyap SK, Maherchandani S, Kumar N. Role of MAPK/MNK1 signaling in virus replication. *Virus Research*. 2018;253:48–61.

Lee GY, Han SN. The role of vitamin E in immunity. *Nutrients*. 2018;10(11):1614.

Li R, Wu K, Li Y, Liang X, Tse WK, Yang L, Lai KP. Revealing the targets and mechanisms of vitamin A in the treatment of COVID-19. *Aging*. 2020;12(15):15784.

Mawson, A.R. 2013. The pathogenesis of malaria: a new perspective. *Pathogens and Global Health*, 107(3), pp. 122–9. doi: 10.1179/2047773213Y.0000000084. PMID: 23683366; PMCID: PMC4003589.

McNally JD, Sampson M, Matheson LA, Hutton B, Little J. Vitamin D receptor (VDR) polymorphisms and severe RSV bronchiolitis: a systematic review and meta-analysis. *Pediatric Pulmonology*. 2014;49:790–9.

Moatasim Y, Kutkat O, Osman AM, Gomaa MR, Okda F, El Sayes M, Kamel MN, Gaballah M, Mostafa A, El-Shesheny R, Kayali G. Potent antiviral activity of vitamin B12 against severe acute respiratory syndrome coronavirus 2, Middle East respiratory syndrome coronavirus, and human coronavirus 229E. *Microorganisms*. 2023;11(11):2777.

Niki E. Role of vitamin E as a lipid-soluble peroxyl radical scavenger: in vitro and in vivo evidence. *Free Radical Biology and Medicine*. 2014;66:3–12.

Omrani AS, Shalhoub S. Middle East respiratory syndrome coronavirus (MERS-CoV): what lessons can we learn? *Journal of Hospital Infection*. 2015;91(3):188–96.

Ramadan N, Shaib H. Middle East respiratory syndrome coronavirus (MERS-CoV): a review. *Germs*. 2019;9(1):35.

Riphagen IJ, Keyzer CA, Drummen NE, De Borst MH, Beulens JW, Gansevoort RT, Geleijnse JM, Muskiet FA, Navis G, Visser ST, Vermeer C. Prevalence and effects of functional vitamin K insufficiency: the PREVEND study. *Nutrients*. 2017;9(12):1334.

Sarohan AR. COVID-19: endogenous retinoic acid theory and retinoic acid depletion syndrome. *Medical Hypotheses*. 2020;144:110250.

Schwalfenberg GK. A review of the critical role of vitamin D in the functioning of the immune system and the clinical implications of vitamin D deficiency. *Molecular Nutrition & Food Research*. 2011;55(1):96–108.

Science M, Maguire JL, Russell ML, Smieja M, Walter SD, Loeb M. Low serum 25-hydroxyvitamin D level and risk of upper respiratory tract infection in children and adolescents. *Clinical Infectious Diseases*. 2013;57(3):392.

Tang N, Li D, Wang X, Sun Z. Abnormal coagulation parameters are associated with poor prognosis in patients with novel coronavirus pneumonia. *Journal of Thrombosis and Haemostasis*. 2020;18(4):844–7.

Timoneda J, Rodríguez-Fernández L, Zaragozá R, Marín MP, Cabezuelo MT, Torres L, Viña JR, Barber T. Vitamin A deficiency and the lung. *Nutrients*. 2018;10(9):1132.

Vyas SP, Hansda AK, Kaplan MH, Goswami R. Calcitriol regulates the differentiation of IL-9-Secreting Th9 cells by modulating the transcription factor PU. 1. *The Journal of Immunology*. 2020;204(5):1201–13.

WHO MERS-CoV Research Group. State of knowledge and data gaps of Middle East respiratory syndrome coronavirus (MERS-CoV) in humans. *PLoS Currents*. 2013;5.

Zhang J, Sun RR, Yan ZX, Yi WX, Yue B. Correlation of serum vitamin A, D, and E with recurrent respiratory infection in children. *European Review for Medical and Pharmacological Sciences*. 2019;23(18):8133–8.

5

Role of lipid-soluble vitamins in MERS-related infections

RAJAN MALHOTRA, GURSEEN RAKHRA, PRIYA CHOUHAN, GURMEEN RAKHRA, AND RATNA RABHA

INTRODUCTION

Vitamins are considered the most vital form of micronutrients which have diversified biochemical roles in the improvement of neurological disorders and the immune system. They are imparted in the form of supplements in diets or food products (Panigrahi et al., 2019). There are four lipid-soluble vitamins, such as vitamins A, D, E, and K, based on their structure (Carlberg, 1999). The supplementation of vitamins for the improvement of innate immunity of the body in patients who were suffering from viral infections had presented positive impacts (Samad et al., 2021). The maintenance of well-proportioned nutritional status in people is the chief method to prevent viral infection. The nutritional references are given to improve the immune system of the body, reduce lung infections, and limit the damage to the lungs due to COVID-19 infection (Samad et al., 2021). These lipid-soluble vitamins (A, D, E, and K) protect the patient's body from getting the infection by activating the defence mechanism and most potentially its other complications like inflammation and cytokine storm. The increase in these two above-stated processes can lead to mortality and morbidity of the patients (Samad et al., 2021). Vitamin D, being a steroidal kind of hormone, works as an immunomodulator by increasing the function of macrophage and reducing the inflammatory cytokine's expression. The protective effects of 1, 25-DHCC (vitamin D) have the potential to reduce the inflammation and increase the immunity of the innate type. Influenza is considered to be a result of vitamin D deficiency. Vitamin A (retinoic acid) in low amounts is supposed to be better than in higher amounts because it can inhibit the growth of the cell and can be cytotoxic, teratogenic, and mutagenic in nature if present in higher concentrations (Mawson, 2013). The functions of lipid-soluble vitamins in the body are depicted in Figure 5.1.

ROLE OF LIPID-SOLUBLE VITAMINS IN MERS AND RELATED INFECTIONS

The first case of Middle East respiratory syndrome (MERS) was described in 2012 in Saudi Arabia (Park et al., 2018). Nearly 10 years later, the appearance of SARS-CoV was noticed due to the death of men, which was initially supposed to be due to renal failure and acute pneumonia. Later on, the MERS virus was isolated using the sputum sample of that patient. More cases were reported in April 2012 in a hospital located in Jordan, which was confirmed as the case of MERS diagnosis (de Wit et al., 2016). The MERS virus is a serious pathogen in humans and is linked with a death rate of 40% in the cases reported (Mackay and Arden, 2015).

Vitamin A has long been known to have immunomodulatory properties and is required for the immune system to function properly. A lack of vitamin A, for example, has been linked to an increased risk of death from respiratory infections. Vitamin A influences a variety of cellular responses to viruses, including cytokine release, acute and chronic inflammatory responses, immunoglobulin synthesis, and a variety of other immunological processes (Mekky et al., 2022). Vitamin D's main role against COVID-19 is that it acts as an immunosuppressant, inhibiting the cytokine release syndrome, which is one of the symptoms of severe COVID-19 (Mekky et al., 2022). Furthermore, vitamin D acts as a mediator for Th2 cell differentiation and the release of their cytokines (IL-4 and IL-10), both of which are important in preventing organ damage

DOI: 10.1201/9781003435686-5

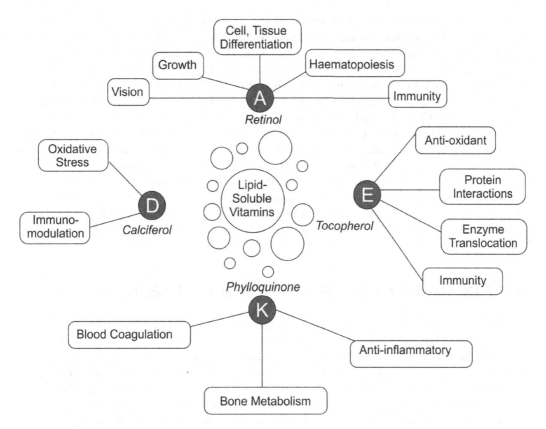

Figure 5.1 A representation of the various functions performed by lipid-soluble vitamins.

caused by severe COVID-19 infection. Furthermore, several clinical trials are being conducted to investigate the role of vitamin D in the prevention and treatment of autoimmune diseases due to its immunomodulatory function. Vitamin D's neuroprotective effect is also important against COVID-19 (Mekky et al., 2022). Vitamin K deficiency and anti-thrombotic activity were linked to decreased activation of endothelial protein S and the presence of other hepatic procoagulant factors and the matrix Gla protein (MGP). A novel relationship between inactive MGP levels and respective vitamin K levels allowed its association with COVID-19, while an increase in active MGP levels reflected the role of vitamin K in inhibiting the coronavirus's lethal effects and inducing elastic fibre damage (Mekky et al., 2022). Erol et al. discovered low vitamin E levels in SARS-CoV-2-infected pregnant women, owing to the virus's oxidative stress that leads to negative perinatal effects. Furthermore, due to its previously mentioned immune-promoting properties, supplementation of vitamin E with another set of nutrients in obese patients has been linked to more positive outcomes in COVID-19 patients.

ROLE OF VITAMIN A IN MERS AND RELATED INFECTIONS

It is a lipid-soluble vitamin, and the human body is not capable of synthesising it via a de novo process, but it is a dietary factor of mandatory type. All-trans-retinoic acid, which binds to particular nuclear transcription factors (retinoid receptors), mediates the actions of vitamin A by controlling the expression of hundreds of genes. Genes that are regulated by vitamin A are involved in basic biological processes, including those that support vision, growth, cell and tissue differentiation, haematopoiesis, and immunity. A reduction in infection-related morbidity and mortality linked to vitamin A deficiency is associated with vitamin A supplementation (Pecora et al., 2020).

Retinoids are closely linked molecules associated with vitamin A that have potent immune-modulating properties and the ability to escalate and increase the efficacy of interferon-I (IFN-I). Furthermore, in the last 60 years, retinoids and related compounds have been well documented as a safe therapeutic option. As a result, numerous studies have reported that retinoids may be useful in the treatment of COVID-19. There are seven central targets that work against COVID-19 and are recently searched out such as MAPK1 (mitogen-activated protein kinase 1), MAPK14 (mitogen-activated protein kinase 14), EGFR (epidermal growth factor receptor), ICAM1 (intercellular adhesion molecule 1), CAT (catalase), PRKCB (protein kinase C-β), and IL-10 (interleukin 10).

Furthermore, vitamin A has been shown to reduce COVID-19-induced adverse effects on the angiotensin system and medication-related side effects. Aside from these novel benefits, vitamin A is well known for its role in promoting innate and adaptive immunity and has been shown to prevent or reduce primary and secondary infections. As a result, it improves respiratory health by lowering inflammation and fibrosis. Furthermore, COVID-19 has been shown to induce inflammatory responses/cytokine storms, particularly in the liver, kidney, and lungs, which increase the risk of depletion of vitamin A storage that requires supplementation and has the potential to restore the acceptable status and combat the severe life-threatening disease (Samad et al., 2021).

ROLE OF VITAMIN D IN MERS AND RELATED INFECTIONS

Monocytes, macrophages, T and B lymphocytes, and other immune cells all have vitamin D receptors. These cells have a 25(OH)D-1-hydroxylase enzyme that changes 25(OH)D into 1,25-dihydroxyvitamin D, which is the active form of vitamin D. Numerous autoimmune and inflammatory diseases have a correlation with vitamin D status. Chronic diseases can be severely worsened by respiratory tract infections, increasing the risk of death. There are various mechanisms by which vitamin D works to reduce the risk of pneumonia and other respiratory infections (Akhtar et al., 2020). Numerous bodily functions, including both innate and adaptive immune responses, depend on vitamin D. Through the stimulation of the expression of antimicrobial peptides like cathelicidin and defensins, vitamin D improves innate cellular immunity. Defensins support adherens, tight and gap junctions, and they promote the expression of genes that fight oxidative stress. The integrity of epithelial tight junctions is known to be significantly damaged by viruses like influenza, which raises the risk of infection and pulmonary oedema. The integrity of these junctions is known to be preserved by vitamin D, and inflammation and increased claudin-2 expression are both caused by low levels of vitamin D receptor expression (Shakoor et al., 2021).

Reduced sunlight exposure and/or preexisting vitamin D deficiency are hypothesised to increase the accumulation, expression, and potential toxicity of endogenous retinoids (i.e. decrease the vitamin D-to-vitamin A ratio), which trigger viral activation or increase host susceptibility to novel influenza virus strains. Furthermore, while normal physiological retinoid concentrations appear to work with vitamin D to inhibit influenza pathogenesis, higher background concentrations (very low vitamin D:A ratios) worsen the disease and may cause severe or fatal complications (Mawson, 2013).

GENERAL FEATURES AND ROLE OF VITAMIN E IN MERS AND CLOSELY RELATED INFECTIONS

Vitamin E functions as an antioxidant being lipid soluble. There are eight different forms of vitamin E: four tocopherols and four tocotrienols. However, human blood and tissues contain the highest concentrations of tocopherol (Beck, 2007). It is found to be localised in the areas such as membrane of organelle and cell membrane where it can exert the highest protective effects of its type, even in the concentration which can be as low as like one vitamin E molecule available for 2000 phospholipid molecules (Bivona et al., 2017).

Because the activity of signalling enzymes is regulated by the redox state, vitamin E's antioxidant activity may be responsible for the regulation of several enzymes involved in signal transduction. Vitamin E inhibits protein kinase C (PKC) activity by increasing PKC dephosphorylation via protein phosphatase 2A activation.

Vitamin E has been shown to inhibit PKC in various cells, resulting in platelet aggregation inhibition; reduced proliferation of monocytes, macrophages, neutrophils, and vascular smooth muscle cells; and decreased super-oxide production in neutrophils and macrophages. Vitamin E can bind directly to enzymes involved in the production of lipid mediators and transport proteins involved in signal transduction. Vitamin E may influence membrane protein interaction and enzyme translocation to the plasma membrane, altering the activity of sig-nal transduction enzymes (Lee and Han, 2018). It has the proficiency to protect unsaturated fatty acids (UFAs) from peroxidation, which are found in the cell membrane; this is the way in which vitamin E does contribute to the stability of cell membrane. The animal studies and the clinical trial too helped to study the advantageous effects on the immune system imparted by the supplementation of vitamin E (Morales-Gonzalez, 2018).

Vitamin E's biological activity is highly reliant on regulatory mechanisms that serve to retain α-tocopherol and excrete non-α-tocopherol forms. This preference is determined by the function of α-tocopherol transfer protein (α-TTP) to enrich plasma with α-tocopherol and the metabolism of non-α-tocopherols. Mutations in this protein cause severe vitamin E deficiency, characterised by neurologic abnormalities, particularly ataxia, and ultimately death if vitamin E is not provided in large quantities to compensate for the lack of α-TTP (Niki and Traber, 2012). There are not many studies done on COVID-19 and related diseases for vitamin E, but in a recent study done on vitamin E, it was found that the number of infections of the respiratory system was expressively low in the group that was given a dose of vitamin E than in the group that didn't supplement with vitamin E (Park et al., 2023).

ASSOCIATION OF VITAMIN K WITH MERS AND RELATED INFECTIONS

Vitamin K is commonly known to be the fat-soluble vitamin, which is established to play a significant role in bone health and blood coagulation. This vitamin is known to exist in the form of menaquinone and phyl-loquinone in nature (Kang et al., 2022). Vitamin K has a core structure of the 2-methyl-1,4-naphthoquinone ring and a side chain at the C3 position of the naphthoquinone ring structure. Menaquinones have a fully unsaturated side chain composed of 2–13 isopentenyl units, whereas phylloquinone has a partially unsatu-rated side chain composed of one isopentenyl followed by three isopentyl units. Menaquinone subtypes with a different number (n) of isoprenoid units are known as menaquinone-n (MK-n). MK-4, for example, has four isoprenoid units on the side chain that are linked to a 2-methyl-1,4-naphthoquinone ring. The structural variation of vitamin K influences biological functions. The isoprene side chain of vitamin K influences its half-life and intestinal absorption (Kang et al., 2022). Compounds with classical vitamin K cofactor activity for the conversion of specific peptide-bound glutamate (Glu) residues to carboxyglutamate (Gla) all share a 2-methyl-1,4-naphthoquinone structure known as menadione and a side chain at the 3-position (Shearer and Newman, 2008). Vitamin K is considered essential for bone metabolism because it participates in the carboxylation of Vitamin K-dependent Proteins (VKDPs) such as osteocalcin (OC), prothrombin, and MGP. To assess the nutritional and clinical significance of vitamin K, serum vitamin K homologue levels or dietary vitamin K intake must be measured. Phylloquinone (vitamin K1) and menaquinones such as MK-4 and MK-7 are vitamin K homologues (Tsugawa and Shiraki, 2020).

Vitamin K deficiency has been linked to vascular calcification and osteoporosis. Vitamin K is also thought to have anti-inflammatory properties, which are mediated by a decrease in circulating inflammatory media-tors. A recent study found that hospitalised adults with COVID-19 infection had low vitamin K levels (Desai et al., 2021).

ROLE OF VITAMINS D AND E IN RELATED INFECTIONS OF MERS

Influenza is a kind of respiratory viral illness, which is caused by three major viral subtypes (A, B, and C) that infect and replicate in human epithelial cells lining the respiratory tract. The most lethal of the three subtypes is influenza A, which is found to be linked with yearly epidemics and infrequent pandemics. Type A has been isolated from humans, birds, pigs, horses, and sea mammals, whereas types B and C have only been isolated from humans. The epidemics of influenza A characteristically affect the group of very young and the group of elder people and have a global reach. The elderly are disproportionately affected by influenza-related deaths, which are primarily caused by pneumonia (Mawson, 2013).

Vitamin D reduces the menace of microbial infection and death through a variety of mechanisms. A recent study on vitamin D's role in lowering the risk of the common cold classified those mechanisms into three types: adaptive immunity, cellular natural immunity, and physical barriers. Vitamin D also boosts cellular immunity by reducing the cytokine storm caused by the innate immune system. In response to viral and bacterial infections, the innate immune system produces both pro-inflammatory and anti-inflammatory cytokines, as seen in COVID-19 patients. Vitamin D has been shown to reduce the production of pro-inflammatory Th1 cytokines like tumour necrosis factor and interferon. Vitamin D administration reduces pro-inflammatory cytokine expression while increasing anti-inflammatory cytokine expression in macrophages (Grant et al., 2020).

Vitamin E deficiency impairs both humoral (antibody production) and cell-mediated (particularly T-cell) immune functions, according to animal and human studies. While vitamin E deficiency impairs immune function, animal and human studies show that supplementing with vitamin E above current dietary recommendations improves immune function. Supplementing with vitamin E improves T-cell-mediated functions in animals, including thymic T-cell differentiation, lymphocyte proliferation, IL-2 production, and helper T-cell activity (Lewis et al., 2019).

VITAMIN A AND ITS ROLE IN MERS AND RELATED DISEASE: PROPHYLAXIS AND GENERAL MECHANISM

MERS-CoV infection spreads from animals to humans and from humans to humans. Evidence suggests that bats aided as the first host species for MERS-CoV. Surprisingly, recent research has revealed that MERS-CoV in bats infects humans. Human–bat interaction or secretion is an occasional transition to "MERS-CoV" and is involved in virus transmission to humans. The most common cause of "MERS-CoV" infections is exposure to animals, specifically bats and "dromedary camels" (Nassar et al., 2018). MERS-CoV has evolved proteins that can effectively counteract human antiviral responses, as have many other viruses (Banerjee et al., 2019). Although MERS and SARS appear clinically similar, in vitro studies have revealed significant differences in these viruses' growth characteristics, receptor usage, and host responses, implying that their pathogenesis may be quite different (van den Brand et al., 2015). Using human tissue that has been infected ex vivo is one way to predict changes in the lungs after MERS-CoV infection. Using spectral confocal microscopy, Hocke et al. demonstrated widespread MERS-CoV antigen expression in type I and II alveolar cells, ciliated bronchial epithelium, and unciliated cuboid cells of terminal bronchioles.

Virus antigen was also discovered in pulmonary vessel endothelial cells, but it is rare to find in alveolar macrophages. Electron microscopy revealed alveolar epithelial damage, including type II alveolar epithelial cell detachment and associated tight junction disruption, chromatin condensation, nuclear fragmentation, and membrane blebbing, the latter indicating apoptosis (van den Brand et al., 2015).

After entering the respiratory tract, the MERS virus primarily interacts with the host DPP4 receptor via its spike (S) protein. DPP4 receptors can be found on the epithelial surface of many human organs, including the lungs, kidneys, liver, bone marrow, thymus, and intestines (Choudhry et al., 2019). Reports of autopsy from fatal MERS-CoV patients revealed that both type I and type II pneumocytes expressed DPP4 and became infected with MERS-CoV, implying that DPP4-expressing type I pneumocytes play a role in MERS-CoV pathogenesis. During viral infection, type I cells in the lung alveoli may be damaged, resulting in diffuse alveolar damage. In line with human MERS cases, common marmosets that express DPP4 in both type I and type II pneumocytes were found to produce more infectious virus following experimental

MERS-CoV infection than rhesus and cynomolgus macaques that only expressed DPP4 in type II pneumocytes (Widagdo et al., 2019).

While MERS-CoV and SARS-CoV have similar pathogenesis and disease aetiology, the molecular mechanisms they use during infection differ. MERS-CoV, for example, has been shown to be significantly more susceptible to type I interferon (IFN) treatment than SARS-CoV; however, research indicates that MERS-CoV has a greater capacity to modulate downstream interferon-stimulated gene responses. Furthermore, SARS-CoV and MERS-CoV have no sequence homology in the context of their accessory open reading frame proteins, implying that the viruses' immune modulation differs. Our previous research had highlighted the importance of SARS-CoV accessory Open Reading Frames (ORFs) in robust infection. Similarly, the functions of accessory ORFs 4a and 4b in modulating aspects of the host response indicate a similar requirement during MERS-CoV infection (Menachery et al., 2017).

The pathogenesis of coronavirus infection begins from mild phases of symptoms and progresses to severe and asymptomatic conditions. As soon as it binds to epithelial cells of the respiratory tract, the virus commences its replication processes and goes down into the epithelial cells of alveoli in the lung region through the passage of airways. During the spread of infection inside the body, fast replication of virus happens that may trigger the immune deregulation and immune responses (Jiang et al., 2022). There are a range of signalling pathways which are involved in the pathogenesis such as Toll Like Receptor (TLR) pathway, IL-1β pathway, NF-κβ pathway, Mitogen Activated Protein (MAP) pathway, and PI3/AKT signalling pathway (Jiang et al., 2022).

TLR pathway: TLRs are innate immune receptors that play an important role in the activation of innate immunity, cytokine expression regulation, indirect activation of the adaptive immune system, and recognition of pathogen-associated molecular patterns (PAMPs). TLR pathways, as a component of innate immunity, could be involved in the pathogenesis of SARS-CoV-2 because previous research has shown that TLRs play an important role in the pathogenesis of SARS-CoV and MERS. According to research, type 1 IFN plays an important role in SARS-CoV and MERS-CoV. They disrupt host cell signalling pathways and reduce the expression of IFN receptors, resulting in a systemic inflammatory response. TLRs may play an important role in the pathogenesis of CoVs because they mediate the production of type 1 IFN. Furthermore, a cytokine storm exists in MERS-CoV, SARS-CoV, and SARS-CoV-2, which plays a critical role in the progression of these infections. TLR3 via the TIRF pathway causes a protective response in SARS-CoV and MERS-CoV infections, according to several studies. Although TLR3 activation of the IRF3 and NF-kB pathways leads to the production of type 1 IFN and pro-inflammatory cytokines in mouse models, there is no reduction in the secretion of these cytokines during coronavirus infection in TLR3 knockout mice. As a result, other pathways play a role in the production of pro-inflammatory cytokines and type 1 IFN. Induction of the TLR3 pathway, but not TLR2/4/7, stimulates IFN production in macrophages and inhibits infection in murine coronavirus infection in mouse models. MERS-CoV 4a protein can also bind to dsRNA and suppress the production of type 1 IFN. TLR4 stimulates the same pathway (Khanmohammadi and Rezaei, 2021).

IL-1β pathway: IL-1 is a pro-inflammatory cytokine that belongs to the IL-1 cytokine family. The cytokine IL-1 is made up of two components: IL-1α and IL-1β. The IL-1α and IL-1β genes are located adjacent to each other on chromosome 2 (2q14.1) and encode IL-1α and IL-1β, respectively. COVID-19 infection causes epithelial injury, which leads to IL-1α secretion (which causes neutrophils and Monocytes (MOs) to be recruited to the infection site) and IL-1β production in MOs/Macrophages (MQs). IL-1β and IL-1Ra have been detected in patients with COVID-19-induced pneumonia's peripheral blood and bronchoalveolar lavage fluid (BALF). Following inflammasome activation, IL-1β, a pro-inflammatory cytokine, is activated and released. Intriguingly, patients with severe/critical COVID-19 infection had significantly higher levels of inflammatory cytokines in their BALFs than patients with mild COVID-19 infection, particularly IL-8, IL-6, and IL-1β (Makaremi et al., 2022).

NF-κB pathway: NF-κB is a transcription factor family that can promote the transcription of stress-response proteins as well as pro-inflammatory cytokines and chemokines. NF-κB activation is critical in the inflammatory response to respiratory viruses like human coronaviruses. Several structural and non-structural proteins of SARS-CoV, including the N protein, S protein, nsp1, nsp3a, and nsp7a, have been shown to stimulate NF-κB activation. In order to respond to SARS-CoV infection, which is a stressful condition, increased secretion of interleukin (IL)-1 is required, which is accomplished via two pathways: cleavage of pro-IL-1 and stimulation of pro-IL-1 transcription. The increased expression of IL-1, in turn, increases the expression of several pro-inflammatory cytokines, including TNF-α and IL-6, which leads to the activation of inflammasomes. ORF4b binds to karyopherin-4 (importin-3) during MERS-CoV infection in Huh-7 and Calu-3 cells, inhibiting its interaction with NF-B-p65 and preventing NF-B nuclear translocation. As a result of the latter event, the expression of pro-inflammatory cytokines and NF-B-dependent cytokines increases (Hemmat et al., 2021).

MAPK pathway: Mitogen-activated protein kinases (MAPKs) are serine/threonine kinases that regulate important cellular functions like cell cycle progression, gene transcription, and post-transcriptional regulation. MAPKs are classified into three groups: p38 family kinase, Jun N-terminal kinase (JNK), and extracellular signal-regulated kinase (ERK). Radiation, oxidative stimulation, inflammatory cytokines, and viral infections can all stimulate the p38 MAPK pathway's expression. The pathogenesis of disease caused by a viral infection involves the activation or inactivation of p38 MAPK. In SARS-CoV-infected cells, the p38 MAPK signalling pathway plays some roles. A recent study showed that p38 MAPK is phosphorylated in SARS-CoV-infected cells, resulting in increased levels of phosphorylated eIF4E (which promotes virus-specific protein synthesis), dephosphorylation of Tyr-705 STAT3, and then inactivation via p38 MAPK activation (which induces apoptotic cell death in SARS-CoV-infected cells) (Farahani et al., 2022).

PI3/AKT signalling pathways: Cell proliferation, antiviral immunity, protein translation, RNA processing, apoptosis, and autophagy are all regulated by the PI3K/AKT pathway. Viruses stimulate and exploit several host cellular signalling pathways to allow successful replication cycles. The PI3K/AKT pathway has recently received a lot of attention due to its role in virus replication control. This pathway has been shown to be necessary not only for viral cell entry but also for intracellular trafficking and replication of some viruses. PI3K/AKT signalling is also involved in the upregulation of the interferon response, and increasing PI3K/AKT activity can stop viral spread by triggering cellular defences. Influenza virus stimulation of this pathway reduces IRF-3-based promoter functions and dimerisation of IRF-3, resulting in a decrease in host antiviral activity (Hemmat et al., 2021).

Vitamin A (VA) is required to keep the respiratory and intestinal epithelial barriers healthy. Although VA deficiency does not cause obvious changes in healthy epithelial surfaces, pathological changes to infected epithelial surfaces were found to be much greater in VA-deficient mice than in control mice following viral infection in the intestine, for example. Following noxious exposures such as ozone treatment and influenza virus infection, VA deficiency increases epithelial damage and impairs recovery, sometimes leading to squamous metaplasia in alveoli and airways. Sputum (mucus) also acts as a physical barrier to pathogens and contains many innate immune system macromolecules. All-trans-retinoic acid regulates mucin production via the major retinoid receptor subtype, retinoic acid receptor alpha (Stephensen and Lietz, n.d.).

Retinol, retinal, and retinoic acid, which each include an alcohol, aldehyde, or carboxylic acid end group, are the three most physiologically active forms of vitamin A (Polcz and Barbul, 2019). Retinol is carried into the bloodstream through transthyretin and retinol-binding protein 4 (RBP4) (prealbumin). Retinol enters cells in its target tissues either passively by diffusion or by activating one of two multi-transmembrane receptors such as retinoic acid 6 (STRA6) or retinol-binding protein receptor 2 (RBPR2). Lecithin:retinol acyltransferase (LRAT) or acyl-CoA:diacylglycerol acyltransferase 1 esterifies retinol when it passes through its receptor and binds to cellular retinol-binding protein 1 (RBP1) (DGAT1). Retinol is sequestered by this esterification, which stops it from entering oxidative metabolic pathways and leaving the cell. Through oxidation/reduction reactions, unesterified retinol can be changed into retinal (VanBuren and Everts, 2022). The time- and dose-dependent effects of morphogenesis, epithelial cell proliferation and differentiation, epithelial and mesenchymal synthetic performance, immune modulation, stimulation of angiogenesis, and inhibition of carcinogenesis are the foundation of the cellular mechanisms of action of natural retinoids (Reichrath et al., 2007). Vitamin A deficiency is found to be associated with the severity of measles, pneumonia, and diarrhoea (Jason et al., 2002). Figure 5.2 shows the involvement of vitamins A, D, E, and K in different mechanisms responsible for their role in viral infections.

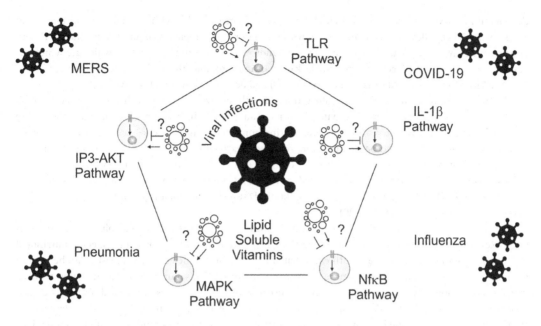

Figure 5.2 An illustration of the viral infections associated with the pathways and vitamins A, D, E, and K.

ROLE OF STEROIDAL VITAMINS IN MERS AND RELATED DISEASES, PROPHYLAXIS, AND THERAPEUTIC INTERVENTIONS

Old-age personnel are more vulnerable to such virus infections because of low and weak immune system power, and hence, when any infection-causing organism from airborne medium comes into their contact, they are usually affected, and most dominantly, they are subjected to vital organ damage like cardiovascular system and lung damage. The intake of vitamin D reduces all the risks related to COVID-19 (Kumar et al., 2021). Vitamin D is a secosteroid and has the immunomodulatory, antioxidant, anti-fibrotic, and anti-inflammatory actions, so it works in a wide background of spectrum. Vitamin D also suppresses the overexpression of cytokine TH1. In adults and old people, the deficiency of vitamin D in their bodies was reported, and it was concluded that these patients are more vulnerable to COVID-19 infection positively (Ebadi and Montano-Loza, 2020). Vitamin D decreased the rate of death when consumed as a supplement. It has a negative impact on HIV and influenza. It is also called the negative modulator of renin–angiotensin system (RAS) (Murdaca et al., 2020). Vitamins A–E are the most suitable against COVID-19-causing infection as an antioxidant, local paracrine signalling, and a natural barrier. Vitamins C and D and thiamine are effective for respiratory diseases – coronavirus infection, which creates havoc in the survival of human life (Jovic et al., 2020).

The SARS-CoV-2 outbreak and rapid spread pose a global health threat with an uncertain outcome. A recent study found that vitamin D has antiviral properties, as well as the ability to inhibit viral replication and be anti-inflammatory and immunomodulatory. Recent reviews have identified some pathways by which vitamin D reduces the risk of microbial infections. Vitamin D works through various mechanisms to reduce the risk of viral infection and mortality. Vitamin D works through three channels to reduce the risk of catching a cold: the physical barrier, cellular natural immunity, and adaptive immunity. A recent review also supported vitamin D's potential role in lowering the risk of COVID-19 infections and mortality. These include cell junction and gap junction maintenance, increasing cellular immunity by decreasing the cytokine storm with influence on interferon-γ and tumour necrosis factor-α, and regulating adaptive immunity by inhibiting T helper cell type 1 responses and stimulating T-cell induction. Vitamin D supplementation has also been shown to increase CD4+ T-cell count in HIV patients. Vitamin D was found to be active in lung tissue and to protect against experimental interstitial pneumonitis in both mouse models and human cell lines. Several in vitro studies have shown that vitamin D plays an important role in local "respiratory homeostasis" by either stimulating the expression of antimicrobial peptides or directly interfering with respiratory virus replication. As a result, vitamin D deficiency may be involved in Acute Respiratory Distress Syndrome (ARDS) and heart failure, which are symptoms of severely ill COVID-19 subjects. As a result, a lack of vitamin D promotes the renin–angiotensin system (RAS), which can lead to chronic cardiovascular disease (CVD) and decreased lung function (Ali, 2020).

The influenza virus infects the respiratory tract directly or by interfering with the immune system's response. The ensuing pneumonia is usually the direct cause of death. Patients with pneumonia are more likely to be 5 years old, over 65 years old, white, nursing home residents, have chronic lung or heart disease, have a history of smoking, and be immunocompromised. Seasonal influenza infections are most common in the winter. The winter peak was thought to be due in part to the season when solar UVB doses, and thus 25(OH)D concentrations, are lowest in most mid- and high-latitude countries. According to ecological studies, increasing 25(OH)D concentrations through vitamin D supplementation in the winter reduces the risk of developing influenza (Grant et al., 2020).

Beta-carotene is the plant-derived precursor to vitamin A, which was the first fat-soluble vitamin identified. Researchers believe that an impaired immune response is caused by a lack of a specific nutritional element. Measles and diarrhoea are both exacerbated by vitamin A deficiency, and measles can be fatal in vitamin A-deficient children. Furthermore, Semba et al. reported that vitamin A supplementation reduced morbidity and mortality in a variety of infectious diseases, including measles, diarrhoeal disease, measles-related pneumonia, HIV infection, and malaria. Low vitamin A diets, according to Jee et al., may reduce the effectiveness of inactivated bovine coronavirus vaccines and make calves more susceptible to infectious disease (Zhang and Liu, 2020).

In chickens fed a diet marginally deficient in vitamin A, the effect of infection with infectious bronchitis virus (IBV), a type of coronavirus, was more pronounced than in those fed a diet adequate in vitamin A. As a result, vitamin A may be a promising treatment option for this novel coronavirus and the prevention of lung infection (Zhang and Liu, 2020).

COVID-19 was discovered in the winter of 2019 and primarily affected middle-aged to elderly people. People who have been infected with the virus may be deficient in vitamin D. Furthermore, decreased vitamin D status in calves has been linked to bovine coronavirus infection. As a result, vitamin D could be used as another therapeutic option for treating this novel virus (Zhang and Liu, 2020).

As an antioxidant, vitamin E plays an important role in reducing oxidative stress by binding to free radicals. Vitamin E deficiency has been shown to worsen the myocardial injury caused by coxsackievirus B3 (a type of RNA virus) infection in mice, as well as to increase the virulence of coxsackievirus B3 in mice due to vitamin E or selenium deficiency. Furthermore, the decreased vitamin E and D status in calves resulted in bovine coronavirus infection (Zhang and Liu, 2020).

NUTRITIONAL SUPPLEMENTATION OF VITAMINS A, D, AND E AS VIRAL PROPHYLACTIC

It is very clear that vitamin D has a relevant role in combating the viral infections. It is cross-checked and found that the people with low concentrations of vitamin D are more vulnerable and always at high risk of spreading infection during COVID-19 time. High doses of D_3 are recommended for all those patients who are already infected with COVID-19. The doses of about 40–60 ng/mL are a dose of recovery for COVID-19 patients (Grant et al., 2020). The spread of viral infection initiates from the trimeric protein present in the spikes on the surface. The spike proteins are responsible for viral attachment to the human host at receptor via ACE2 which as done by evolutionary-related counterpart SARS-CoV-1 (Zabetakis et al., 2020). The overall role of fat-soluble vitamins as a treatment for SARS-CoV-2 infection came into existence during pandemic situations when no possible treatment was available. The supplementation of vitamin D is crucial still when other vitamins are most important because it is only vitamin D which can synthesise the whole variety of vitamins by itself. Medical nutritional therapy is an immune-based therapy in which all malnutrition is considered to be essential to fulfil the body's need to protect from such infections (Samad et al., 2021).

Vitamin A is another alternative for immunomodulation. It is stored in the liver in hepatic stellate cells, and it gets hydrolysed in retinol from retinyl esters. It has an effective depleting effect on RNA-based viruses and helps to combat diseases spread by viruses like Ebola and measles. It has minimal side effects and works synergistically when vitamin A supplements with vaccines are provided. It maintains an immune response that can help in the recovery of COVID-19 patients. It is nowadays on testing to elute out its therapeutic role to overcome COVID-19 illness in patients (Midha et al., 2021).

A clear demonstration from a number of studies has revealed that the clinical importance of supplements in the form of diet has a great therapeutic importance at the time of pandemic. The impact of vitamins on respiration-related illnesses like pneumonia, common cold, and influenza has shown a deliberate success to defend from virus-causing threat of survival. The diet, supplements, glutathione, and other vitamins have a potentially different role in the prevention of viral infections (Singh et al., 2022).

The role of diet and supplements nowadays is not only limited to health, but it also provides abundant strength and a potential to fight and withstand the situation of infection spreading and causing threat to the survival of beings. The invasion of microbes can be minimised when an intake of potentially powerful functional foods, probiotics, nutraceuticals, multi-nutrient supplements, and vitamins is taken in everyday diet as a therapeutic way to combat infections. A novel antiviral drug can be a milestone to treat such stubborn viruses which cause adverse havoc in the human world (Singh et al., 2021).

The initial intake of micronutrients before undergoing any medications with a purpose of treating outbreak-causing SARS-CoV-2 is a preventive measure for combating such disease spreading. The primary symptom that appears is inflammation of the respiratory tract. So, to reduce inflammation and boost immune response, it is a primary need to continue intake, essentially supply of nutrients. There is always a close association between nutrition and COVID-19 invasion in a healthy individual (Mohammadi et al., 2023).

CHALLENGES AND DOSE-RELATED ILL-EFFECTS OF VITAMINS A AND D IN MERS

The deficiency of vitamin A is a result of nutrition depletion. It is a cause of severe respiratory disorder in children and infants who are at great risk of viral infections of SARS-CoV and MERS-CoV. An effective preparation against viral infection – amphiphilic polyanhydride nanoparticle-based vaccines – was developed. The mechanism through which viral infection enters the respiratory and mucosal defence system was also recorded (McGill et al., 2019). There is a link between mortality and deficiency of vitamin D. It also depends on sex, age, and gender of those who have taken doses of vitamin supplements to comorbidities so that the severity can be overcome (Radujkovic et al., 2020). Targeted therapy is the only possible treatment for such respiratory disorders in which antibodies are introduced in the host to fight against and obstruct viral introduction in the body, inhibit viral replication, and ultimately control virion formation from proteases. Attenuated viral vaccines and peptide-based vaccines are the best options to clear cut the interactions between host and responses of the immune system (Marian, 2021). The preparations that act naturally are the only way in which we can dehumidify, detoxify, and moisturise the lungs so that these unfavourable conditions will not allow viruses to invade the host (Sun et al., 2022). To overcome the deficiency of vitamin A in a large proportion of the population considering women and children worldwide, it is necessary to increase the intake of vitamin A in their diet. The effect of therapeutic doses of micronutrients on lactating mothers, as their energy demand is more than others, is enough to protect them from many diseases (Penniston and Tanumihardjo, 2006). The effects of vitamin A on the immune system are to prevent the potential growth of carriers who have adverse effects on the human body and are a cause of viral diseases such as measles, malaria, diarrhoea, and HIV. It increases lymphopoiesis of the T-cell. It also improves the antibody responses to the vaccines prepared for the treatment of measles, diphtheria toxoid, and tetanus. Vitamin A is obtained from sources of animals in the form of retinyl palmitate (Villamor and Fawzi, 2005). Vitamin D is an immunomodulator and responsible for other essential functions like stimulation of pro-inflammatory cytokine secretion, and augmenting the barrier functions by neutrophil activity. Vitamin D also has a booster role for innate responses of immunity. It obstructs viral introduction and gives a new direction for the adaptive immune response for more synthesis of T helper cell type 2. It is recommended that the intake of vitamin D during COVID-19 time is the best remedy for patients suffering from COVID-19 (Rawat et al., 2021). The low level of vitamin D indicates the severity of infection susceptibility to overcome such a MERS-CoV-2 infection (Gönen et al., 2021). In recent years, gold nanoparticles are being studied in an early pandemic against MERS-CoV to develop a label-free colorimetric assay (Kim et al., 2019). After post-mortem analysis, it was reported that for those who were suffering from MERS-CoV-2, the level of angiotensin-converting enzyme-2 was relatively low in their lung tissues. It is an innovative therapy to block ACE2 to not let S protein of SARS and MERS bind specifically to the receptor and is used as an approach for the treatment or prevention of COVID-19 (Datta et al., 2020). Convalescent therapy (CP) is another approach for the prevention or treatment of MERS. It is related to adaptive immunity where immunotherapy of plasma is given to the patients who are thought to be suffering from COVID-19 (Khulood et al., 2020).

REFERENCES

Akhtar, S., Das, J.K., Ismail, T., Wahid, M., Saeed, W., Bhutta, Z.A., 2020. Nutritional perspectives for the prevention and mitigation of COVID-19. *Nutr Rev.* https://doi.org/10.1093/nutrit/nuaa063

Ali, N., 2020. Role of vitamin D in preventing of COVID-19 infection, progression and severity. *J Infect Public Health* 13, 1373–1380. https://doi.org/10.1016/j.jiph.2020.06.021

Banerjee, A., Baid, K., Mossman, K., 2019. Molecular pathogenesis of Middle East Respiratory Syndrome (MERS) coronavirus. *Curr Clin Micro Rpt* 6, 139–147. https://doi.org/10.1007/s40588-019-00122-7

Beck, M.A., 2007. Selenium and vitamin E status: impact on viral pathogenicity. *J Nutr* 137, 1338–1340. https://doi.org/10.1093/jn/137.5.1338

Bivona, J.J., Patel, S., Vajdy, M., 2017. Induction of cellular and molecular immunomodulatory pathways by vitamin E and vitamin C. *Expert Opin Biol Ther* 17, 1539–1551. https://doi.org/10.1080/14712598.2017.1375096

Carlberg, C., 1999. Lipid soluble vitamins in gene regulation. *BioFactors* 10, 91–97. https://doi.org/10.1002/biof.5520100202

Choudhry, H., Bakhrebah, M.A., Abdulaal, W.H., Zamzami, M.A., Baothman, O.A., Hassan, M.A., Zeyadi, M., Helmi, N., Alzahrani, F., Ali, A., Zakaria, M.K., Kamal, M.A., Warsi, M.K., Ahmed, F., Rasool, M., Jamal, M.S., 2019. Middle East respiratory syndrome: pathogenesis and therapeutic developments. *Future Virol* 14, 237–246. https://doi.org/10.2217/fvl-2018-0201

Datta, P.K., Liu, F., Fischer, T., Rappaport, J., Qin, X., 2020. SARS-CoV-2 pandemic and research gaps: understanding SARS-CoV-2 interaction with the ACE2 receptor and implications for therapy. *Theranostics* 10, 7448–7464. https://doi.org/10.7150/thno.48076

de Wit, E., van Doremalen, N., Falzarano, D., Munster, V.J., 2016. SARS and MERS: recent insights into emerging coronaviruses. *Nat Rev Microbiol* 14, 523–534. https://doi.org/10.1038/nrmicro.2016.81

Desai, A.P., Dirajlal-Fargo, S., Durieux, J.C., Tribout, H., Labbato, D., McComsey, G.A., 2021. Vitamin K & D deficiencies are independently associated with COVID-19 disease severity. *Open Forum Infect Dis* 8, ofab408. https://doi.org/10.1093/ofid/ofab408

Ebadi, M., Montano-Loza, A.J., 2020. Perspective: improving vitamin D status in the management of COVID-19. *Eur J Clin Nutr* 74, 856–859. https://doi.org/10.1038/s41430-020-0661-0

Farahani, M., Niknam, Z., Mohammadi Amirabad, L., Amiri-Dashatan, N., Koushki, M., Nemati, M., Danesh Pouya, F., Rezaei-Tavirani, M., Rasmi, Y., Tayebi, L., 2022. Molecular pathways involved in COVID-19 and potential pathway-based therapeutic targets. *Biomed Pharmacother* 145, 112420. https://doi.org/10.1016/j.biopha.2021.112420

Gönen, M.S., Alaylıoğlu, M., Durcan, E., Özdemir, Y., Şahin, S., Konukoğlu, D., Nohut, O.K., Ürkmez, S., Küçükece, B., Balkan, İ.İ., Kara, H.V., Börekçi, Ş., Özkaya, H., Kutlubay, Z., Dikmen, Y., Keskindemirci, Y., Karras, S.N., Annweiler, C., Gezen-Ak, D., Dursun, E., 2021. Rapid and effective vitamin D supplementation may present better clinical outcomes in COVID-19 (SARS-CoV-2) patients by altering serum INOS1, IL1B, IFNg, Cathelicidin-LL37, and ICAM1. *Nutrients* 13, 4047. https://doi.org/10.3390/nu13114047

Grant, W.B., Lahore, H., McDonnell, S.L., Baggerly, C.A., French, C.B., Aliano, J.L., Bhattoa, H.P., 2020. Evidence that vitamin D supplementation could reduce risk of influenza and COVID-19 infections and deaths. *Nutrients* 12, 988. https://doi.org/10.3390/nu12040988

Hemmat, N., Asadzadeh, Z., Ahangar, N.K., Alemohammad, H., Najafzadeh, B., Derakhshani, A., Baghbanzadeh, A., Baghi, H.B., Javadrashid, D., Najafi, S., Ar Gouilh, M., Baradaran, B., 2021. The roles of signaling pathways in SARS-CoV-2 infection; lessons learned from SARS-CoV and MERS-CoV. *Arch Virol* 166, 675–696. https://doi.org/10.1007/s00705-021-04958-7

Jason, J., Archibald, L.K., Nwanyanwu, O.C., Sowell, A.L., Buchanan, I., Larned, J., Bell, M., Kazembe, P.N., Dobbie, H., Jarvis, W.R., 2002. Vitamin A levels and immunity in humans. *Clin Vaccine Immunol* 9, 616–621. https://doi.org/10.1128/CDLI.9.3.616-621.2002

Jiang, Y., Zhao, T., Zhou, X., Xiang, Y., Gutierrez-Castrellon, P., Ma, X., 2022. Inflammatory pathways in COVID-19: mechanism and therapeutic interventions. *MedComm* (2020) 3, e154. https://doi.org/10.1002/mco2.154

Jovic, T.H., Ali, S.R., Ibrahim, N., Jessop, Z.M., Tarassoli, S.P., Dobbs, T.D., Holford, P., Thornton, C.A., Whitaker, I.S., 2020. Could vitamins help in the fight against COVID-19? *Nutrients* 12, 2550. https://doi.org/10.3390/nu12092550

Kang, M.-J., Baek, K.-R., Lee, Y.-R., Kim, G.-H., Seo, S.-O., 2022. Production of vitamin K by wild-type and engineered microorganisms. *Microorganisms* 10, 554. https://doi.org/10.3390/microorganisms10030554

Khanmohammadi, S., Rezaei, N., 2021. Role of toll-like receptors in the pathogenesis of COVID-19. *J Med Virol* 93, 2735–2739. https://doi.org/10.1002/jmv.26826

Khulood, D., Adil, M.S., Sultana, R., Nimra, null, 2020. Convalescent plasma appears efficacious and safe in COVID-19. *Ther Adv Infect Dis* 7, 2049936120957931. https://doi.org/10.1177/2049936120957931

Kim, H., Park, M., Hwang, J., Kim, J.H., Chung, D.-R., Lee, K., Kang, M., 2019. Correction to Development of label-free colorimetric assay for MERS-CoV using gold nanoparticles. *ACS Sens* 4, 2554–2554. https://doi.org/10.1021/acssensors.9b01450

Kumar, R., Rathi, H., Haq, A., Wimalawansa, S.J., Sharma, A., 2021. Putative roles of vitamin D in modulating immune response and immunopathology associated with COVID-19. *Virus Res* 292, 198235. https://doi.org/10.1016/j.virusres.2020.198235

Lee, G.Y., Han, S.N., 2018. The role of vitamin E in immunity. *Nutrients* 10, 1614. https://doi.org/10.3390/nu10111614

Lewis, E.D., Meydani, S.N., Wu, D., 2019. Regulatory role of vitamin E in the immune system and inflammation. *IUBMB Life* 71, 487–494. https://doi.org/10.1002/iub.1976

Mackay, I.M., Arden, K.E., 2015. MERS coronavirus: diagnostics, epidemiology and transmission. *Virol J* 12, 222. https://doi.org/10.1186/s12985-015-0439-5

Makaremi, S., Asgarzadeh, A., Kianfar, H., Mohammadnia, A., Asghariazar, V., Safarzadeh, E., 2022. The role of IL-1 family of cytokines and receptors in pathogenesis of COVID-19. *Inflamm Res* 71, 923–947. https://doi.org/10.1007/s00011-022-01596-w

Marian, A.J., 2021. Current state of vaccine development and targeted therapies for COVID-19: impact of basic science discoveries. *Cardiovasc Pathol* 50, 107278. https://doi.org/10.1016/j.carpath.2020.107278

Mawson, A.R., 2013. Role of fat-soluble vitamins A and D in the pathogenesis of influenza: a new perspective. *ISRN Infect Dis* 2013, 1–26. https://doi.org/10.5402/2013/246737

McGill, J.L., Kelly, S.M., Guerra-Maupome, M., Winkley, E., Henningson, J., Narasimhan, B., Sacco, R.E., 2019. Vitamin A deficiency impairs the immune response to intranasal vaccination and RSV infection in neonatal calves. *Sci Rep* 9, 15157. https://doi.org/10.1038/s41598-019-51684-x

Mekky, R.Y., Elemam, N.M., Eltahtawy, O., Zeinelabdeen, Y., Youness, R.A., 2022. Evaluating risk: benefit ratio of fat-soluble vitamin supplementation to SARS-CoV-2-infected autoimmune and cancer patients: do vitamin-drug interactions exist? *Life* 12, 1654. https://doi.org/10.3390/life12101654

Menachery, V.D., Mitchell, H.D., Cockrell, A.S., Gralinski, L.E., Yount, B.L., Graham, R.L., McAnarney, E.T., Douglas, M.G., Scobey, T., Beall, A., Dinnon, K., Kocher, J.F., Hale, A.E., Stratton, K.G., Waters, K.M., Baric, R.S., 2017. MERS-CoV accessory ORFs play key role for infection and pathogenesis. *mBio* 8, e00665-17. https://doi.org/10.1128/mBio.00665-17

Midha, I.K., Kumar, N., Kumar, A., Madan, T., 2021. Mega doses of retinol: a possible immunomodulation in Covid-19 illness in resource-limited settings. *Rev Med Virol* 31, e2204. https://doi.org/10.1002/rmv.2204

Mohammadi, A.H., Behjati, M., Karami, M., Abari, A.H., Sobhani-Nasab, A., Rourani, H.A., Hazrati, E., Mirghazanfari, S.M., Hadi, V., Hadi, S., Milajerdi, A., 2023. An overview on role of nutrition on COVID-19 immunity: accumulative review from available studies. *Clin Nutr Open Sci* 47, 6–43. https://doi.org/10.1016/j.nutos.2022.11.001

Morales-Gonzalez, J. A. (Ed.). (2018). *Vitamin E in Health and Disease*. InTech. doi: 10.5772/intechopen.74469

Murdaca, G., Pioggia, G., Negrini, S., 2020. Vitamin D and Covid-19: an update on evidence and potential therapeutic implications. *Clin Mol Allergy* 18, 23. https://doi.org/10.1186/s12948-020-00139-0

Nassar, M.S., Bakhrebah, M.A., Meo, S.A., Alsuabeyl, M.S., Zaher, W.A., 2018. Middle East Respiratory Syndrome Coronavirus (MERS-CoV) infection: epidemiology, pathogenesis and clinical characteristics. *Eur Rev Med Pharmacol Sci* 22(15), 4956–4961.

Niki, E., Traber, M.G., 2012. A history of vitamin E. *Ann Nutr Metab* 61, 207–212. https://doi.org/10.1159/000343106

Panigrahi, S.S., Syed, I., Sivabalan, S., Sarkar, P., 2019. Nanoencapsulation strategies for lipid-soluble vitamins. *Chem Pap* 73, 1–16. https://doi.org/10.1007/s11696-018-0559-7

Park, J.-E., Jung, S., Kim, A., Park, J.-E., 2018. MERS transmission and risk factors: a systematic review. *BMC Public Health* 18, 574. https://doi.org/10.1186/s12889-018-5484-8

Park, J.-H., Lee, Y., Choi, M., Park, E., 2023. The role of some vitamins in respiratory-related viral infections: a narrative review. *Clini Nutr Res* 12, 77–89. https://doi.org/10.7762/cnr.2023.12.1.77

Pecora, F., Persico, F., Argentiero, A., Neglia, C., Esposito, S., 2020. The role of micronutrients in support of the immune response against viral infections. *Nutrients* 12, 3198. https://doi.org/10.3390/nu12103198

Penniston, K.L., Tanumihardjo, S.A., 2006. The acute and chronic toxic effects of vitamin A231-4. *Am J Clin Nutr* 83, 191–201. https://doi.org/10.1093/ajcn/83.2.191

Polcz, M.E., Barbul, A., 2019. The role of vitamin A in wound healing. *Nutr Clin Pract* 34, 695–700. https://doi.org/10.1002/ncp.10376

Radujkovic, A., Hippchen, T., Tiwari-Heckler, S., Dreher, S., Boxberger, M., Merle, U., 2020. Vitamin D deficiency and outcome of COVID-19 patients. *Nutrients* 12, 2757. https://doi.org/10.3390/nu12092757

Rawat, D., Roy, A., Maitra, S., Shankar, V., Khanna, P., Baidya, D.K., 2021. Vitamin D supplementation and COVID-19 treatment: a systematic review and meta-analysis. *Diabetes Metab Syndr* 15, 102189. https://doi.org/10.1016/j.dsx.2021.102189

Reichrath, J., Lehmann, B., Carlberg, C., Varani, J., Zouboulis, C.C., 2007. Vitamins as hormones. *Horm Metab Res* 39, 71–84. https://doi.org/10.1055/s-2007-958715

Samad, N., Dutta, S., Sodunke, T.E., Fairuz, A., Sapkota, A., Miftah, Z.F., Jahan, I., Sharma, P., Abubakar, A.R., Rowaiye, A.B., Oli, A.N., Charan, J., Islam, S., Haque, M., 2021. Fat-soluble vitamins and the current global pandemic of COVID-19: evidence-Based efficacy from literature review. *J Inflamm Res* 14, 2091–2110. https://doi.org/10.2147/JIR.S307333

Shakoor, H., Feehan, J., Al Dhaheri, A.S., Ali, H.I., Platat, C., Ismail, L.C., Apostolopoulos, V., Stojanovska, L., 2021. Immune-boosting role of vitamins D, C, E, zinc, selenium and omega-3 fatty acids: could they help against COVID-19? *Maturitas* 143, 1–9. https://doi.org/10.1016/j.maturitas.2020.08.003

Shearer, M.J., Newman, P., 2008. Metabolism and cell biology of vitamin K. *Thromb Haemost* 100, 530–547. https://doi.org/10.1160/TH08-03-0147

Singh, B., Eshaghian, E., Chuang, J., Covasa, M., 2022. Do diet and dietary supplements mitigate clinical outcomes in COVID-19? *Nutrients* 14, 1909. https://doi.org/10.3390/nu14091909

Singh, S., Kola, P., Kaur, D., Singla, G., Mishra, V., Panesar, P.S., Mallikarjunan, K., Krishania, M., 2021. Therapeutic potential of nutraceuticals and dietary supplements in the prevention of viral diseases: a review. *Front Nutr* 8, 679312. https://doi.org/10.3389/fnut.2021.679312

Stephensen, C.B., Lietz, G., n.d. Vitamin A in resistance to and recovery from infection: relevance to SARS-CoV2. *Br J Nutr* 1–10. https://doi.org/10.1017/S0007114521000246

Sun, Y., An, X., Jin, D., Duan, L., Zhang, Yuehong, Yang, C., Duan, Y., Zhou, R., Zhao, Y., Zhang, Yuqing, Kang, X., Jiang, L., Lian, F., 2022. Model exploration for discovering COVID-19 targeted traditional Chinese medicine. *Heliyon* 8, e12333. https://doi.org/10.1016/j.heliyon.2022.e12333

Tsugawa, N., Shiraki, M., 2020. Vitamin K nutrition and bone health. *Nutrients* 12, 1909. https://doi.org/10.3390/nu12071909

van den Brand, J.M., Smits, S.L., Haagmans, B.L., 2015. Pathogenesis of Middle East respiratory syndrome coronavirus. *J Pathol* 235, 175–184. https://doi.org/10.1002/path.4458

VanBuren, C.A., Everts, H.B., 2022. Vitamin A in skin and hair: an update. *Nutrients* 14, 2952. https://doi.org/10.3390/nu14142952

Villamor, E., Fawzi, W.W., 2005. Effects of vitamin a supplementation on immune responses and correlation with clinical outcomes. *Clin Microbiol Rev* 18, 446–464. https://doi.org/10.1128/CMR.18.3.446-464.2005

Widagdo, W., Sooksawasdi Na Ayudhya, S., Hundie, G.B., Haagmans, B.L., 2019. Host determinants of MERS-CoV transmission and pathogenesis. *Viruses* 11, 280. https://doi.org/10.3390/v11030280

Zabetakis, I., Lordan, R., Norton, C., Tsoupras, A., 2020. COVID-19: the inflammation link and the role of nutrition in potential mitigation. *Nutrients* 12, 1466. https://doi.org/10.3390/nu12051466

Zhang, L., Liu, Y., 2020. Potential interventions for novel coronavirus in China: a systematic review. *J Med Virol* 92, 479–490. https://doi.org/10.1002/jmv.25707

6

The role of vitamins in influenza A and related infections
Prophylactic and therapeutic benefits

NASREEN AKHTAR AND HIMMAT SINGH

INTRODUCTION

"I had a little bird
Its name was Enza
I opened the window
And in flew Enza."

(Children's marching rhyme, 1918 Ann Tatlock, *The Names of the Stars*)

Influenza is notable for its health and economic impact (Coker & Mounier-Jack, 2006; De Francisco et al., 2015). Influenza viruses pose a significant global threat to public health, with acute respiratory infections (ARI) being a major burden on global disease. It occurs in two epidemiological forms: epidemics and pandemics. Annual epidemics of seasonal influenza result in 3–5 million cases of severe illness and between 250,000 and 500,000 deaths worldwide (WHO Report on Influenza Virus 2024). The World Health Organization (WHO) estimates that each year, influenza infects approximately 5%–10% of adults and 20%–30% of children (WHO Report on Influenza Virus 2024). According to Ritchie and Roser (2018), Acute Respiratory Infection (ARI) was the third leading cause of death worldwide in 2017. These infections are primarily caused by RNA viruses, although bacteria can also contribute. Influenza pandemics are infrequent compared with seasonal influenza, but they are widely feared public health emergencies because they entail serious social disruption and substantial economic cost (Social and Economic Impact of Influenza, 2024). It is important to note that these pandemics can affect not only high-risk groups but also healthy young individuals who are typically less affected by seasonal flu.

Since its founding in 1948, the WHO has recognized its responsibility to develop and update strategies to prevent and control both pandemics and seasonal influenza through national programs and prepare globally (WHO Influenza Pandemic Plan, 1999). Although the pandemic is global, WHO encourages Member States to develop their national influenza programs based on WHO guidelines for influenza pandemic preparedness and control, including a list of planning checks that include the essential and desirable elements of pandemic preparedness. Nearly 100 countries have used their national plans to respond to the 2009 H1N1 flu pandemic. Only a few countries, mainly European countries and one Southeast Asian country, Thailand, have updated their plan based on the experience of the 2009 H1N1 influenza pandemic. This review focuses primarily on assessing the impact of pharmaceutical, medical, and non-medical interventions. On May 16, 2009, India's first pandemic influenza case was detected in Hyderabad (Kulkarni et al., 2014; Ministry of Health and Family Welfare Report on Influenza A virus). A month later, a software engineer returning from the United States to Pune was admitted to the Infectious Diseases Hospital in Pune. On June 22, 2009, he became the first imported case of H1N1 influenza in Pune (Pune's first case of swine flu, a techie, Indian Express, 2009). Thirty sporadic cases of H1N1 infection were diagnosed by the end of July 2009, but transmission was not widespread (Purohit et al., 2018). An outbreak at a school in the densely populated western suburbs of Pune received extensive media coverage during this period.

DOI: 10.1201/9781003435686-6

India's first pandemic-related death occurred at a private hospital in Pune on August 3, 2009 (Tandale et al., 2010). Community spread of infection soon followed, followed by panic due to lack of information to competent authorities about the risk of infection. The deceased's family has accused the hospital of negligence and filed a complaint with the consumer court. Newspapers covered the case extensively. This made private doctors reluctant to treat patients with flu-like symptoms, and after August 4, long lines of symptomatic patients lined up at three public hospitals in Pune (https://indianexpress.com/article/cities/pune/day-after-rush-at-naidu-hospital-no-need-to-panic-says-niv-director/). Within a week, facing a shortage of test kits, these hospitals were no longer able to test and accept patients.

Influenza B infections are exclusive to humans, while influenza A viruses can infect both pigs and horses. Wild birds, particularly waterfowl, are considered a reservoir for the influenza A virus, as all HA and NA types can be found in them (Webster, 1998; Webster, 2023). Highly pathogenic subtypes H5 and H7 can cause infections in chickens and other bird species, leading to the clinical presentation of classical avian plague, commonly known as bird flu. The reason behind this is that the virus can replicate throughout the bird's body, resulting in a disease with a high mortality rate. Since 1977, there has been a co-circulation of influenza A/H3N2 and A/H1N1 subtypes, as well as influenza B, in humans. Sporadic cases of human viruses belonging to the H1N2 subtype were first observed in 1977, but they were not reported in subsequent years until the 2001/2002 season when they caused localized outbreaks in some countries. However, these viruses did not gain epidemiological significance and have not been identified since 2005 (Gregory et al., 2002).

The human influenza virus is prevalent worldwide, with seasonal epidemics occurring in both the Northern and Southern hemispheres during winter. Interestingly, the winter outbreak in the Southern hemisphere coincides with the summer in the Northern hemisphere due to the seasonal shift. These influenza epidemics are responsible for an estimated 500,000 deaths annually across the globe (Fauci, 2006). While limited information is available regarding the epidemiology of influenza in tropical countries, it is believed that influenza can occur throughout the year in these regions. Despite the clear correlation with seasons, influenza infections can occur outside of the typical influenza epidemics in respective countries, leading to localized and temporary outbreaks. During these annual epidemics, it is estimated that 10%–15% of the population is affected. The severity of the epidemics can vary, allowing for clear distinctions between them. In Germany, national influenza surveillance relies on three main pillars: The National Reference Centre for Influenza at the Robert Koch Institute, the Influenza Working Party's sentinel system, and the mandatory reporting of virus identification by diagnostic laboratories. In recent years, the influenza virus activity was exceptionally high during the 2002/2003 and 2004/2005 seasons. From 2000/2001 to 2004/2005, these annual outbreaks resulted in approximately 2,900,000 additional consultations, 900,000 sick notes among the working population (aged 16–60), 14,000 extra hospitalizations, and around 10,000 deaths compared to periods without influenza virus circulation. The majority of influenza-related deaths occur among individuals over the age of 60, not only in Germany but also in other countries (Zucs et al., 2005; Arbeitsgemeinschaft Influenza (AGI): Saison-berichte; Thompson et al., 2003).

Influenza pandemics are characterized by the occurrence (recurrence) of influenza A subtype against which the majority of the human population is not immune, thus causing a worldwide epidemic. The past century was characterized by three major pandemics: The Spanish Flu of 1918 (H1N1) caused about 40 million deaths, while the death rates of the Asian influenza of 1957 (H2N2) and the Hong Kong Flu of 1968 (H2N3) were estimated at 1–2 million and 0.75–1 million deaths, respectively (Fauci, 2006). The outbreak of classical avian influenza in the Netherlands and Belgium in the spring of 2003, caused by influenza A H7N7, and the subsequent influenza A H5N1 epidemic in South-East Asia starting in 1998 garnered significant public attention due to their impact on humans. There is concern that the next pandemic could also be caused by the highly pathogenic H5N1 subtype of the influenza virus. So far, this virus has been primarily found in (wild) birds, with only isolated cases of human infection. Since the first human H5N1 outbreaks in 2003, the WHO has officially confirmed 331 cases of infection and 202 deaths as of October 12, 2007. Notably, the age group most affected by the virus has been adolescents and young adults. Currently, two areas, Indonesia and Egypt, have experienced recurrent human infections and deaths. The possibility of a genetic predisposition has been discussed, as outbreaks within families have primarily affected close relatives.

INFLUENZA (FLU): SURVEILLANCE AND TRIVIAL PROPHYLAXIS

Influenza is an infectious respiratory disease caused by influenza viruses that infect the respiratory tract. The flu can range from mild to severe and in some cases can even lead to death. Although anyone can get the flu, people 65 years and older, young children, and people with other health conditions are at higher risk of hospitalization and flu-related complications. The best way to reduce your risk of getting the flu and its complications is to get vaccinated every year. Three types of human influenza viruses have been recognized (types A, B, and C), distinguished by their antigenically distinct internal proteins: nucleoproteins and matrix proteins. These viruses are classified as members of the genus *Orthomyxovirus* in the family Orthomyxoviridae and are named according to their type, location where isolated, the successive isolate number from that location, and year of isolation. Only influenza A and B cause large epidemics and severe diseases; influenza C is associated with cold-like illnesses, mainly in children. It is an enveloped virus that varies in size from 80 to 120 nm; its genome consists of eight single-stranded negative-sense RNA segments coding for 10 or 11 proteins, depending on the isolate. The virus has a spherical structure containing RNA material, which is attached to two surface glycoproteins, hemagglutinin and neuraminidase. Fifteen different types of hemagglutinin (H) and nine different types of neuraminidase (N) have been recognized (Figure 6.1). Hemagglutinin is involved in the binding of the virus to host cell receptors and in the fusion between the virus and the cell membrane, leading to the release of virion components into the cell. Neuraminidase releases newly produced virions from host cells, facilitates virus spread to target cells in the respiratory tract, and facilitates virus release from infected cells (Figure 6.1).

Global influenza surveillance has been conducted through WHO's Global Influenza Surveillance and Response System (GISRS) since 1952. GISRS is a system fostering global confidence and trust for over half a century, through effective collaboration and sharing of viruses, data, and benefits based on Member States' commitment to a global public health model (Figure 6.2).

The mission of GISRS is to protect people from the threat of influenza by continuously functioning as a:

- global mechanism of surveillance, preparedness, and response for seasonal, pandemic, and zoonotic influenza;
- global platform for monitoring influenza epidemiology and disease; and
- global alert for novel influenza viruses and other respiratory pathogens.

In India, integrated surveillance of influenza-like illness (ILI) and severe acute respiratory illness (SARI) for the detection of human influenza virus and SARS-CoV-2 virus is ongoing through structured ILI/SARI

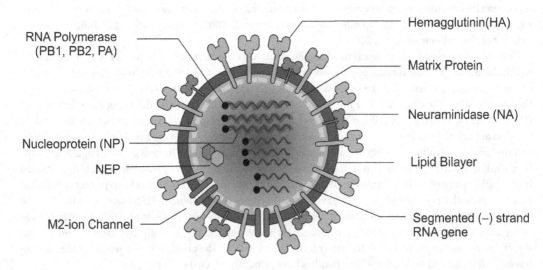

Figure 6.1 Structure of influenza vairus.

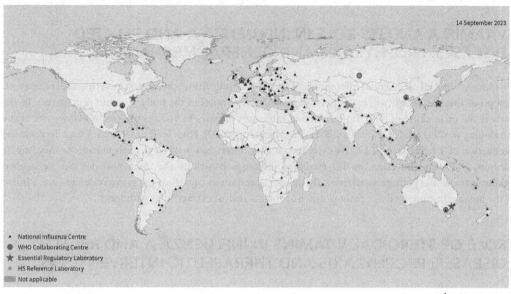

Figure 6.2 WHO Global Influenza Surveillance and Response System (GISRS). (Courtesy: WHO influenza surveillance system. https://www.who.int/initiatives/global-influenza-surveillance-and-response-system.)

surveillance network of 28 sites. The surveillance network is comprised of 27 DHR-ICMR's Virus Research and Diagnostic Laboratories and country's National Influenza Centre (WHO-NIC) housed at ICMR-National Institute of Virology Pune, also a WHO Collaborating Centre for GISRS.

During the period of the first 9 weeks (January 2 to March 5) of 2023, while this book was being written, the surveillance network has monitored the human influenza virus and SARS-CoV-2 infection in Severe Acute Respiratory Illness (SARI) and Influenza like Illness (ILI) cases. The influenza typing results are summarized above (Table 6.1):

Table 6.1 The result of influenza virus from January 2, 2023, to March 5, 2023

Week	Week 1	Week 2	Week 3	Week 4	Week 5	Week 6	Week 7	Week 8	Week 9
Influenza A H1N1pdm09	8	8	4	6	5	3	0	2	5
Influenza A H3N2	46	57	44	42	47	61	46	52	56
Influenza B Victoria	4	11	6	4	12	18	10	13	13

Vaccination is the primary strategy for the prevention and control of influenza (Cox & Subbarao, 1999; Nichol & Treanor, 2006). Although both inactivated vaccines and the live attenuated vaccine are effective in preventing influenza and its associated complications, the protection they confer varies widely, depending on the antigenic match between the viruses in the vaccine and those that are circulating during a given influenza season and on the recipient's age and health status (Fiore et al., 2010). More effective vaccination options are needed, especially for people who have a reduced immunologic response to vaccination, including the elderly and those with chronic underlying disease. The efficacy of the seasonal influenza vaccine ranges between 10%

and 60%. The lowest efficacy occurs when vaccine strains are not well matched to circulating strains. Both the trivalent and quadrivalent vaccines are FDA-approved (Grohskopf et al., 2018).

VITAMIN A AND ITS ROLE IN INFLUENZA A AND RELATED DISEASES: PROPHYLAXIS AND GENERAL MECHANISMS

Vitamin A (Vit A) is an essential micronutrient for maintaining vision, promoting growth and development, and protecting the integrity of epithelial and mucous membranes in the body. Vitamin A deficiency in the body is the main cause of high mortality from infectious diseases in young children in many parts of the developing world (Fawzi et al., 1993; Glasziou & Mackerras, 1993). High level of vitamin A may decrease the production of T helper type-1 (Th1)4 cytokines such as interferon-γ (IFN-γ). Such reductions may impair recovery from viral infection, as Th1-mediated responses constitute the main host defense mechanism against intracellular pathogens, whereas Th1-mediated responses constitute the primary defense mechanism of the host against intracellular pathogens. Th2 protects against extracellular pathogens.

ROLE OF STEROIDAL VITAMINS IN INFLUENZA A AND RELATED DISEASES: PROPHYLAXIS AND THERAPEUTIC INTERVENTIONS

The link between vitamin D deficiency and susceptibility to infection has been suggested for longer than a century, with the early observation that children with nutritional rickets were more likely to experience infections of the respiratory system, leading to the coining of the phrase "rachitic lung" (Khajavi & Amirhakimi, 1977). The pioneering work by Rook (Rook et al., 1986; Crowle et al., 1987) in the 1980s demonstrated that vitamin D enhanced bactericidal activity of human macrophages against *Mycobacterium tuberculosis*, the causative agent of tuberculosis (TB). This discovery led to a new era of interest regarding the role of vitamin D in determining the pathogenesis and the immune response to pathogens. However, the literature and clinical outcomes on roles of vitamin D in influenza, particularly H1N1 is scanty and needs to further evaluated.

NUTRITIONAL SUPPLEMENTATION OF VITAMINS A, D, AND E AS INFLUENZA PROPHYLACTIC

Seasonal influenza infections generally peak in the winter (Hope, 1981). In addition, the winter peak of influenza also coincides with weather conditions of low temperature and relative humidity that allow the influenza virus to survive longer outside the body than under warmer conditions (Lowen et al., 2007; Shaman & Kohn, 2009; Shaman et al., 2010). Vitamins A and D have interactive roles in influenza, and retinoid (the collective term for vitamin A and its natural and synthetic congeners) has an independent role in influenza infection and pathogenesis. For instance, solar radiation has opposite effects on vitamins A and D, catabolizing vitamin A but increasing the concentration of vitamin D; the effects of the two vitamins are mutually inhibitory; retinoids regulate airway epithelial cell growth, differentiation, and gene expression; the symptoms of influenza are similar to those of retinoid toxicity; supplementary and/or pharmacological concentrations of retinoids induce influenza-like symptoms; viral activity is regulated in part by retinoids; and retinoids influence the mechanisms that both inhibit and contribute to influenza pathogenesis. It is hypothesized that reduced sunlight exposure and/or preexisting vitamin D deficiency simultaneously increase the accumulation, expression, and potential toxicity of endogenous retinoids (*i.e.*, decrease the vitamin D-to-vitamin A ratio), which trigger viral activation or increase host susceptibility to novel strains of influenza virus. Furthermore, normal physiological concentrations of retinoid appear to work with vitamin D to inhibit influenza pathogenesis (Mawson, 2013).

Vitamin E is a fat-soluble antioxidant that can protect the polyunsaturated fatty acids (PUFAs) in the membrane from oxidation, regulate the production of reactive oxygen species (ROS) and reactive nitrogen species

(RNS), and modulate signal transduction. Immunomodulatory effects of vitamin E have been observed in animal and human models under normal and disease conditions (Lee & Han, 2018). Vitamin E enhanced immune responses in animal and human models and conferred protection against several infectious diseases. The mechanisms involved with protection against infectious agents were increased macrophage activity and antibody (Ab) production for *Diplococcus pneumoniae* type 1 (Heinzerling et al., 1974) and higher natural killer (NK) activity and Th1 response for influenza virus (Hayek et al., 1997; Han et al., 2000).

CHALLENGES AND DOSE-RELATED ILL EFFECTS OF VITAMINS A AND D IN INFLUENZA A

Influenza infection occurs in people of all ages, but complications are more frequent in elderly people (Dushoff, 2005; Liu et al., 2012). This is partly due to immune dysfunctions caused by aging, *i.e.*, immunosenescence, which can be explained by increased antigenic challenges and chronic inflammation and worsened by deficiencies in nutrients such as vitamin D (Vit D) (Vasson et al., 2013; Pera et al., 2015; Simon et al., 2015). Vitamin D deficiency occurs more frequently in older adults than in young ones because of their lower endogenous vitamin D synthesis and because of their often reduced dietary intake (Castetbon et al., 2009). Vitamin D is known to shift the T-cell response from a T helper 1 (Th1) to a Th2-mediated cell response, thereby reducing inflammation and promoting an immunosuppressive state (Cantorna et al., 2008; Hansdottir et al., 2010; Di Rosa et al., 2011). The public health strategy for influenza is to reduce severe outcomes such as hospitalization and death by recommending annual vaccinations, particularly for people over 65 years old (Trucchi et al., 2015; Grohskopf, 2016). However, the vaccine efficacy is lower for older people (17%–53%) than for young adults (70%–90%) (Goodwin et al., 2006; Jackson et al., 2006).

CONCLUSION

The 1918 influenza pandemic in Spain had a significant impact on global mortality rates, accounting for approximately one-fifth of total deaths worldwide. This devastating pandemic caused widespread havoc and disrupted lives and the economy. Surprisingly, it has received minimal attention from researchers (Arnold, 2019). However, it is important to note that the impact of the pandemic varied greatly across different provinces in India. In the state of Mysore, the well-organized administrative machinery and the existing health and sanitation infrastructure played a crucial role in minimizing the calamity. The chief secretary's instructions and the guidelines provided by the senior surgeon and sanitary commissioner served as a general framework for implementing relief measures. Additionally, the financial sensibilities and beliefs of the masses were taken into consideration. The Mysore administration implemented stern measures to regulate the price of food grains and ensured the free supply of essential goods and medicines, which helped overcome a famine-like situation (Sekher, 2018). However, it is worth noting that the relief measures were not as successful in many rural areas. Nonetheless, these concerns were openly discussed in assembly meetings and reported in newspapers, and the princely government was receptive to these complaints. The century-old experience of Princely Mysore in combating the influenza pandemic provides valuable lessons. These lessons involve a combination of strong administrative measures, such as strict monitoring of public health and sanitation services, timely collection of data and information, well-organized relief operations, regulation of food grain prices, the administration's responsiveness to public grievances and cultural sentiments, and the involvement of civilian and community organizations. Considering the lessons from history on administrative limitations and supply chain cripples, vitamins and their supplementation could prove and effective measure in such crisis. Vitamins, may be considered as an essential emergency commodity in relief measures and components of rapid-relief kits during such pandemics. It is however, required, that various medical and regulatory agencies conduct field studies and trials on requirement, counter-effects and supply chain managements for future implementations.

REFERENCES

Arbeitsgemeinschaft Influenza (AGI): Saison-berichte, Jan 2024. www.influenza.rki.de/agi

Arnold D. Death and the modern empire: The 1918–19 influenza epidemic in India. *Trans R Hist Soc.* 2019, 29:181–200.

Cantorna MT, Yu S, Bruce D. The paradoxical effects of vitamin D on type 1 mediated immunity. *Mol Aspects Med.* 2008, 29:369–375. 10.1016/j.mam.2008.04.004.

Castetbon K, Vernay M, Malon A, Salanave B, Deschamps V, Roudier C, et al. Dietary intake, physical activity and nutritional status in adults: The French nutrition and health survey (ENNS, 2006–2007). *Br J Nutr.* 2009, 102:733–743. 10.1017/S0007114509274745.

Coker R, Mounier-Jack S. Pandemic influenza preparedness in the Asia-Pacific region. *Lancet.* 2006,368(9538):886–889. doi: 10.1016/S0140-6736(06)69209-X

Cox NJ, Subbarao K. Influenza. *Lancet* 1999, 354:1277–1282.

Crowle AJ, Ross EJ, May MH. Inhibition by 1,25(OH)2-vitamin D3 of the multiplication of virulent tubercle bacilli in cultured human macrophages. *Infect Immun.* 1987, 55:2945–2950.

Day after, rush at Naidu Hospital; no need to panic, says NIV director. *Indian Express.* August 5, 2009. https://archive.indianexpress.com/news/day-after-rush-at-naidu-hospital--no-need-t/498107/. (Accessed on January 31, 2024).

De Francisco N, Donadel M, Jit M, Hutubessy R. A systematic review of the social and economic burden of influenza in low-and middle-income countries. *Vaccine.* 2015, 33(48):6537–6544. doi: 10.1016/j.vaccine.2015.10.066.

Di Rosa M, Malaguarnera M, Nicoletti F, Malaguarnera L. Vitamin D3: A helpful immuno-modulator. *Immunology* 2011, 134:123–139. 10.1111/j.1365-2567.2011. 03482.x

Dushoff J. Mortality due to Influenza in the United States-an annualized regression approach using multiple-cause mortality data. *Am J Epidemiol.* 2005, 163:181–187. 10.1093/aje/kwj024.

Fauci AS. Seasonal and pandemic influenza preparedness: science and countermeasures. *J Infect Dis.* 2006, 194(suppl 2):S73–S76.

Fawzi WW, Chalmers TC, Herrera MG, Mosteller F. Vitamin A supplementation and child mortality. A meta-analysis. *J Am Med Assoc.* 1993, 269:898–903.

Fiore AE, Uyeki TM, Broder K, et al. Prevention and control of influenza with vaccines: recommendations of the Advisory Committee on Immunization Practices (ACIP). *MMWR Recomm Rep.* 2010, 59:1–62.

Glasziou PP, Mackerras DE. Vitamin A supplementation in infectious diseases: a meta-analysis. *Br Med J Clin Res.* 1993, 306:366–370.

Goodwin K, Viboud C, Simonsen L. Antibody response to influenza vaccination in the elderly: a quantitative review. *Vaccine* 2006, 24:1159–1169. 10.1016/j.vaccine.2005.08.105.

Gregory V, Bennett M, Orkhan MH, Al Hajjar S, Varsano N, Mendelson E, Zambon M, Ellis J, Hay A, Lin YP. Emergence of influenza A (H1N2) reassortant viruses in the human population during 2001. *Virology.* 2002, 300:1–7.

Grohskopf LA. Prevention and control of seasonal influenza with vaccines. *MMWR Recomm Rep.* 2016, 65:1–54. 10.15585/mmwr. rr6505a1.

Grohskopf LA, Sokolow LZ, Broder KR, Walter EB, Fry AM, Jernigan DB. Prevention and control of seasonal influenza with vaccines: recommendations of the Advisory Committee on Immunization Practices-United States, 2018–19 influenza season. *MMWR Recomm Rep.* 2018, 67(3):1–20.

Han SN, Wu D, Ha WK, Beharka A, Smith DE, Bender BS, Meydani SN. Vitamin E supplementation increases T helper 1 cytokine production in old mice infected with influenza virus. *Immunology.* 2000, 100:487–493. doi: 10.1046/j.1365-2567.2000.00070. x.

Hansdottir S, Monick MM, Lovan N, Powers L, Gerke A, Hunninghake GW. Vitamin D decreases respiratory syncytial virus induction of NF-kappa B-linked chemokines and cytokines in airway epithelium while maintaining the antiviral state. *J Immunol.* 2010, 184:965–974. 10.4049/jimmunol.0902840.

Hayek MG, Taylor SF, Bender BS, Han SN, Meydani M, Smith DE, Eghtesada S, Meydani SN. Vitamin E supplementation decreases lung virus titers in mice infected with influenza. *J Infect Dis*. 1997, 176:273–276. doi: 10.1086/517265.

Heinzerling RH, Tengerdy RP, Wick LL, Lueker DC. Vitamin E protects mice against *Diplococcus pneumoniae* type I infection. *Infect Immun*. 1974, 10:1292–1295.

Hope-Simpson RE. The role of season in the epidemiology of influenza. *J Hyg*. 1981, 86:35–47.doi: 10.1017/s0022172400068728

Influenza (Seasonal). WHO website. https://www.who.int/mediacentre/factsheets/fs211/en/. (Accessed January 30, 2024).

Jackson LA, Jackson ML, Nelson JC, Neuzil KM, Weiss NS. Evidence of bias in estimates of influenza vaccine effectiveness in seniors. *Int J Epidemiol*. 2006, 35(2):337–344. doi: 10.1093/ije/dyi274. Epub 2005 Dec 20. PMID: 16368725.

Khajavi A, Amirhakimi GH. The rachitic lung: pulmonary findings in 30 infants and children with malnutritional rickets. *Clin Pediatr (Phila)* 1977, 16:36–38.

Kulkarni PS, Agarkhedkar S, Lalwani S, Bavdekar AR, Jog S, Raut SK, Parulekar V, Agarkhedkar SS, Palkar S, Mangrule S. Effectiveness of an Indian-made attenuated influenza A (H1N1) pdm 2009 vaccine: a case control study. *Hum Vaccin Immunother*. 2014, 10(3):566–571.

Lee GY, Han SN. The role of vitamin E in immunity. *Nutrients* 2018, 10(11):1614. doi:10.3390/nu10111614

Liu WM, Nahar TER, Jacobi RHJ, Gijzen K, van Beek J, Hak E, et al. Impaired production of TNF-α by dendritic cells of older adults leads to a lower CD8+ T cell response against influenza. *Vaccine*. 2012, 30:1659–1666. 10.1016/j.vaccine.2011.12.105.

Lowen AC, Mubareka S, Steel J, Palese P. Influenza virus transmission is dependent on relative humidity and temperature. *PLoS Pathog*. 2007, 3:1470–1476.

Mawson AR. Role of fat-soluble vitamins A and D in the pathogenesis of influenza: a new perspective. *ISRN Infect Dis*. 2013, 1–26. https://www.researchgate.net/publication/258404997_Role_of_Fat-Soluble_Vitamins_A_and_D_in_the_Pathogenesis_of_Influenza_A_New_Perspective#fullTextFileContent

Nichol KL, Treanor JJ. Vaccines for seasonal and pandemic influenza. *J Infect Dis*. 2006, 194(Suppl 2):S111–S118.

Pera A, Campos C, López N, Hassouneh F, Alonso C, Tarazona R, et al. Immunosenescence: implications for response to infection and vaccination in older people. *Maturitas* 2015, 82:50–55. 10.1016/j.maturitas.2015.05.004.

Pune's first case of swine flu, a techie. *Indian Express*. June 23, 2009. https://archive.indianexpress.com/news/pune-s-first-case-of-swine-flu-a-techie/479906/. (Accessed January 31, 2024).

Purohit V, Kudale A, Sundaram N, Joseph S, Schaetti C, Weiss MG. Public health policy and experience of the 2009 H1N1 influenza pandemic in Pune, India. *Int J Health Policy Manag*. 2018, 7(2):154–166. doi: 10.15171/ijhpm.2017.54. PMID: 29524939; PMCID: PMC5819375.

Ritchie H, Roser M. Causes of Death In: *Our world in data*, 2018.https://ourworldindata.org/causes-of-death

Rook GA, Steele J, Fraher L, et al. Vitamin D3, gamma interferon, and control of proliferation of *Mycobacterium tuberculosis* by human monocytes. *Immunology*. 1986, 57:159–163.

Sekher TV. Addressing public health and sanitation in Mysore, 1881–1921: 'Model' state and 'Native' administrators, In: Ernst W, Pati B and Sekher TV (eds), *Health and Medicine in the Indian Princely States: 1850–1950*, London and New York: Routledge, 2018, pp. 26–41.

Shaman J, Kohn M. Absolute humidity modulates influenza survival, transmission, and seasonality. *Proc Natl Acad Sci USA* 2009, 106:3243–3248.

Shaman J, Pitzer, VE, Viboud, C, Grenfell, BT, Lipsitch, M. Absolute humidity and the seasonal onset of influenza in the continental United States. *PLoS Biol*. 2010, 8:e1000316.

Simon AK, Hollander GA, McMichael A. Evolution of the immune system in humans from infancy to old age. *Proc R Soc B Biol Sci*. 2015, 282:20143085. 10.1098/rspb.2014.3085.

Social and Economic Impact of Influenza - Article in Motion. https://www.rapidreferenceinfluenza.com/chapter/B978-0-7234-3433-7.50013-X/aim/impact-of-pandemic-influenza. Accessed January 30, 2024.

Tandale BV, Pawar SD, Gurav YK, et al. Seroepidemiology of pandemic influenza A (H1N1) 2009 virus infections in Pune, India. *BMC Infect Dis*. 2010, 10:255. doi: 10.1186/1471-2334-10-255.

Thompson W W, Shay DK, Weintraub E, Brammer L, Cox N, Anderson LJ, Fukuda K. Mortality associated with influenza and respiratory syncytial virus in the United States. *JAMA*. 2003, 289:179–186.

Trucchi C Paganino C, Orsi A, De Florentiis D, Ansaldi F. Influenza vaccination in the elderly: why are the overall benefits still hotly debated? *J Prev Med Hyg*. 2015, 56:37-43. doi: 10.15167/2421-4248/jpmh2015.56.1.474.

Vasson M-P, Farges M-C, Goncalves-Mendes N, Talvas J, Ribalta J, Winklhofer-Roob B, et al. Does aging affect the immune status? A comparative analysis in 300 healthy volunteers from France, Austria and Spain. *Immun Ageing A*. 2013, 10:10–38. 10.1186/1742-4933-10-38.

Webster RG. Influenza: an emerging disease. *Emerg Infect Dis*. 1998, 4(3):436–441. doi: 10.3201/eid0403.980325.

Webster RG. Influenza: searching for pandemic origins. *Annu Rev Virol*. 2023, 10(1):1–23. doi: 10.1146/annurev-virology-111821-125223.

World Health Organization (WHO). *Influenza Pandemic Plan. The Role of WHO and Guidelines for National and Regional Planning*. Geneva: WHO; 1999.

World Health Organization (WHO). Influenza (Seasonal) Geneva, Switzerland: World Health Organization, 2020. https://www.who.int/news-room/fact-sheets/detail/influenza-(seasonal). (Accessed January 30, 2024).

World Health Organization (WHO). Global Influenza Surveillance and Response System (GISRS). https://www.who.int/initiatives/global-influenza-surveillance-and-response-system (Accessed on January 31, 2024)

Zucs P, Buchholz U, Haas W, Uphoff H. Influenza associated excess mortality in Germany, 1985–2001. *Emerg Themes Epidemiol*. 2005, 2:6.

Prophylactic and therapeutic roles of water-soluble vitamins in COVID-19

VIPIN KUMAR SHARMA, JYOTI SANGWAN, AHANA DASGUPTA, AND SHILPA RAINA

INTRODUCTION

The onset of the COVID-19 pandemic, instigated by the emergence of the novel coronavirus SARS-CoV-2 in December 2019 in Wuhan, China, swiftly escalated into a global health crisis by March 2020, primarily due to its exceptional transmissibility via respiratory droplets. Examining the historical context of coronaviruses yields valuable insight into their evolutionary trajectory and pathogenicity. SARS-CoV-2 is one of the most dreadful viruses faced by mankind, which not only led to the COVID-19 outbreak in China but also spread throughout the world, infecting more than 528 million people and causing more than 6.3 million deaths worldwide [1,2]. The exploration of human coronaviruses (HCoVs) dates back to the 1960s when researchers Bynoe and Tyrrell identified a virus, later named B814, from a patient with a common cold. Simultaneously, Procknow isolated a virus exhibiting distinct properties in tissue culture. Electron microscopy studies by Tyrrell and Almeida revealed medium-sized, membrane-coated, pleomorphic agents with characteristic crown-like projections, thus categorizing them as coronaviruses [3]. Research expanded to examine the epidemiology and pathogenicity of HCoVs across various animal species [4]. The emergence of SARS-CoV in 2002–2003 and MERS-CoV in 2012 emphasized the zoonotic origins of these viruses, originating in bats and adapting to intermediary hosts such as palm civets and camels before spilling over to humans [5]. These outbreaks, alongside the identification of additional HCoVs like HCoV-NL63 and HCoV-HKU1, underscored the significance of coronaviruses as causative agents of respiratory illnesses affecting individuals across all age groups [6]. Efforts to mitigate the COVID-19 pandemic have been multifaceted, with a primary emphasis on expedited vaccine development and deployment. Several vaccines have attained emergency use authorization, presenting promising avenues for curbing viral transmission and lessening disease burden. Nonetheless, challenges persist, including the emergence of novel variants with heightened transmissibility or resistance to existing vaccines, necessitating continuous surveillance and research into enhanced vaccination strategies.

Preventive measures and therapeutic strategies are indispensable in combating the COVID-19 pandemic. Preventive measures such as mask wearing, hand hygiene, social distancing, and vaccination play a pivotal role in curbing the spread of the virus, thereby reducing the burden on healthcare systems and protecting vulnerable populations. By breaking the chain of transmission, these measures not only help prevent individuals from contracting the virus but also help prevent severe illness, hospitalizations, and deaths. Moreover, they contribute to the overall containment of the pandemic, allowing for the gradual reopening of economies and the restoration of societal activities. Concurrently, therapeutic strategies are crucial for managing COVID-19 cases effectively. Therapeutic interventions such as antiviral drugs, corticosteroids, and monoclonal antibodies help alleviate symptoms, prevent disease progression, and reduce the risk of severe complications in infected individuals [7]. By targeting the virus directly or modulating the immune response, these treatments improve patient outcomes and reduce the strain on healthcare resources. Additionally, ongoing research and development efforts are essential for identifying novel therapeutics and refining existing treatment protocols to enhance patient care and support the global fight against COVID-19.

Nutrition plays a crucial role in supporting immune function, as various nutrients are essential for the proper functioning of the immune system. Adequate intake of vitamins, minerals, antioxidants, and other bioactive compounds is necessary to maintain the body's defense mechanisms against pathogens [8].

For instance, vitamin C, vitamin D, zinc, and selenium are known to have immunomodulatory effects and contribute to the normal functioning of the immune system [9]. These nutrients play roles in the proliferation and activity of immune cells, the production of antibodies, and the regulation of inflammatory responses. Moreover, a well-balanced diet rich in fruits, vegetables, whole grains, lean proteins, and healthy fats provides a wide array of nutrients that support immune health. Phytochemicals such as flavonoids, polyphenols, and carotenoids found in plant-based foods have antioxidant and anti-inflammatory properties that help protect against oxidative stress and chronic inflammation, which can impair immune function [10]. Additionally, omega-3 fatty acids, particularly those found in fatty fish and flaxseeds, have anti-inflammatory effects that may enhance immune function and reduce the risk of inflammatory diseases [11]. Furthermore, malnutrition, including both undernutrition and overnutrition, can compromise immune function and increase susceptibility to infections. Poor dietary habits, such as consuming excessive amounts of processed foods, sugary beverages, and unhealthy fats, can lead to obesity, metabolic syndrome, and other chronic conditions that negatively impact immune health. Therefore, promoting a nutrient-rich diet and healthy eating habits is essential for supporting immune function and reducing the risk of infectious diseases.

Water-soluble vitamins are a group of organic compounds that are essential for various physiological functions in the body and are soluble in water. These vitamins cannot be synthesized in sufficient quantities by the body and must be obtained through the diet. In our body, water-soluble vitamins are vitamin B and vitamin C. These water-soluble vitamins are not stored in the body to a significant extent, and excess amounts are typically excreted in the urine. Therefore, they need to be consumed regularly through a balanced diet to maintain optimal health and prevent deficiencies [10]. Vitamin B, encompassing various B vitamins such as B6, B12, and folate, is essential for immune function and energy metabolism. Vitamin B6, for instance, plays a crucial role in immune cell proliferation and cytokine production, key processes in mounting an effective immune response against viral infections. Vitamin B12 and folate are involved in DNA synthesis and methylation, processes important for proper immune cell function and antibody production. Deficiencies in these B vitamins can compromise immune function and increase susceptibility to infections. In the context of COVID-19, maintaining adequate levels of B vitamins may support immune responses and help mitigate the risk of severe illness [12].

Vitamin C is an indispensable water-soluble nutrient and antioxidant known to bolster the immune system. Various aspects of both innate and adaptive immunity benefit from the presence of vitamin C, thereby enhancing overall immune function. Extensive literature highlights the antioxidant, anti-inflammatory, anticoagulant, and immunomodulatory properties associated with vitamin C. Studies indicate that pneumonia and sepsis patients often exhibit depleted levels of ascorbic acid and heightened oxidative stress [12].

COVID-19 PATHOPHYSIOLOGY AND WATER-SOLUBLE VITAMINS

The unique structure of the SARS-CoV-2 virus showcases particle sizes ranging from 70 to 90 nm, intricately observed within various intracellular organelles, particularly vesicles [13]. There is a notable resemblance in the sequence between SARS-CoV-2 and SARS-CoV, suggesting a similar structural framework [14]. The surface proteins of the virus, such as the spike protein, membrane protein, and envelope protein, are enclosed within a lipid bilayer derived from the host membrane, encapsulating the helical nucleocapsid housing viral RNA [15]. The genomic size of the coronavirus falls within the range of 26–32 kb and consists of 6–11 open reading frames encoding polyproteins containing 9680 amino acids [16]. The first open reading frame encodes 16 nonstructural proteins, while the remaining frames encode both accessory and structural proteins. The genome of SARS-CoV-2 lacks the hemagglutinin–esterase gene but includes two flanking untranslated regions at its ends [17]. Moreover, comparative analysis between SARS-CoV-2 and SARS-CoV reveals minimal differences in open reading frames and nonstructural proteins. The nonstructural proteins include viral cysteine proteases like papain-like protease (nsp3), chymotrypsin-like protease (nsp5), RNA-dependent RNA polymerase (nsp12), helicase (nsp13), and other proteins involved in transcription and replication processes [17]. Additionally, key structural proteins like the spike surface glycoprotein (S), membrane protein, nucleocapsid protein (N), and envelope protein (E) play vital roles in virus morphogenesis and assembly. The spike glycoprotein structure of SARS-CoV-2 closely resembles that of SARS-CoV, with some structural divergence observed in the receptor-binding region. The S protein is vital for cell attachment and virus-to-host membrane fusion, featuring a trimeric structure with distinct RNA binding domains (RBDs) and C-Terminal Domains (CTDs) in the S1 and S2 subunits, respectively [18] (discussed and illustrated in greater detail in Chapter 8).

MECHANISMS OF VIRAL INFECTION AND REPLICATION

Both SARS-CoV-2 and SARS-CoV utilize the angiotensin-converting enzyme 2 (ACE2) receptor to enter host cells, whereas MERS-CoV relies on dipeptidyl peptidase 4 (DPP4) as its host receptor. The entry of coronaviruses into host cells hinges on binding the spike glycoprotein to the cellular receptor and the priming of the spike protein by host cell proteases. Like SARS-CoV, SARS-CoV-2 employs the ACE2 receptor for internalization and TMPRSS2 serine proteases for spike protein priming. The widespread tissue expression of ACE2 receptors contributes to the potential extrapulmonary spread of SARS-CoV-2 [19]. Additionally, studies have revealed that the spike protein of SARS-CoV-2 exhibits significantly higher affinity than that of SARS-CoV [20]. Upon binding to the ACE2 receptor, conformational changes occur in the spike protein, leading to fusion between the viral envelope and the host cell membrane via the endosomal pathway [21]. Proteolytic activity is crucial for exposing the fusion peptide within the spike protein, facilitating fusion with the host membrane. Subsequently, the S protein is cleaved by host cell proteases into S1 and S2 subunits, mediated by transmembrane protease serine 2 (TMPRSS2) and human airway trypsin-like protease (HAT). The S2 subunit comprises a fusion peptide and heptad repeat domains, initiating conformational changes that allow viral entry into the host cell. Following entry, viral RNA is released into the host cytoplasm, which undergoes translation to generate replicase polyproteins. These polyproteins are then cleaved by virus-encoded proteinases into smaller proteins. The replication process involves ribosomal frameshifting during translation, generating both genomic and subgenomic RNA species encoding viral proteins. Assembly of new virions occurs at cellular organelles such as the endoplasmic reticulum (ER) and the Golgi complex, culminating in the release of virions from the host cell via vesicles. The newly formed virions can infect nearby healthy cells and spread to others through highly contagious respiratory droplets, potentially transmitting the disease to uninfected individuals [13,19,22] (Figure 7.1).

VITAMIN B COMPLEX AND COVID-19

Vitamins B6, B12, and folate are integral components of the immune system, contributing to its normal function, as recognized by health claims in the European Union. Deficiencies in these vitamins can have profound impacts on immune responses. Vitamin B6, for instance, is involved in various immune processes, including the function and proliferation of T lymphocytes, which are crucial for orchestrating adaptive immune responses [10]. Additionally, it plays a role in modulating cytokine and chemokine release, key signaling molecules that regulate immune cell communication and inflammation. Folate, also known as vitamin B9, is essential for DNA synthesis and cell division, processes critical for the proliferation of immune cells. Folate deficiency can lead to megaloblastic anemia, characterized by large, immature red blood cells, as well as impairments in immune function. It has been associated with combined immunodeficiency, affecting T-cell proliferation and altering the production of pro-inflammatory cytokines, which are important for mounting effective immune responses against pathogens. Vitamin B12, or cobalamin, is essential for the maintenance of nerve cells and red blood cell formation, but it also plays a role in immune function. In addition to its impact on hematopoiesis, studies have suggested a role for vitamin B12 in mediating immune responses to viral infections, as evidenced by improved outcomes in patients with chronic hepatitis C following supplementation [23].

Moving on to respiratory diseases, thiamine, also known as vitamin B1, is crucial for aerobic respiration as it serves as a cofactor for enzymes involved in energy metabolism. Insufficient thiamine levels can lead to impaired aerobic respiration, forcing cells to rely more on anaerobic pathways, which can result in elevated lactate levels. Thiamine is also involved in antioxidant pathways, alongside niacin, contributing to the production

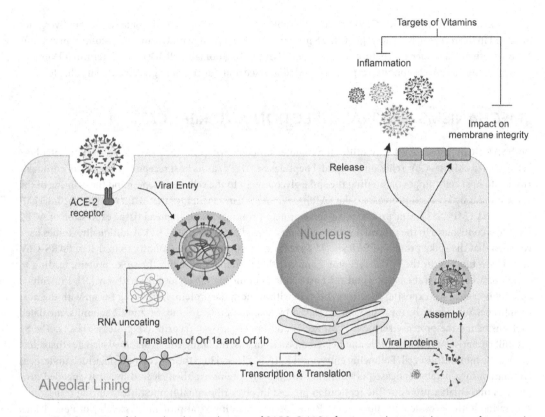

Figure 7.1 Overview of the molecular mechanism of SARS-CoV-2 infection and potential targets of water-soluble vitamins.

of nicotinamide adenine dinucleotide phosphate (NADPH) and glutathione cycling, which are important for neutralizing reactive oxygen species and maintaining cellular homeostasis [12]. Administration of thiamine has shown promising results in critically ill patients, particularly those with septic shock, where it has been associated with reduced lactate levels and improved mortality. Combination therapies involving thiamine, vitamin C, and hydrocortisone have demonstrated significant improvements in organ injury, shock reversal time, and mortality rates in patients with severe sepsis. While studies have reported lower levels of folate and vitamin B12 in patients with chronic obstructive pulmonary disease (COPD), evidence regarding the efficacy of supplementation in improving symptoms, hospitalization rates, or pulmonary function remains limited [11].

EVIDENCE FROM STUDIES ON VITAMIN B SUPPLEMENTATION IN COVID-19

The coronavirus polyprotein encodes two proteases, M-pro and PL-pro, targeted for drug discovery due to their roles in viral replication. A recent study utilized the crystal structure of SARS-CoV-2 M-pro to screen existing drugs for potential repurposing against COVID-19. Vitamin B12 and nicotinamide were identified as potential candidates based on their binding interactions with M-pro. Another computational study highlighted folate's potential to form strong hydrogen bonds with active site residues of M-pro, suggesting it as a therapeutic strategy. These computational screening tools offer opportunities for targeted drug testing using cell-based assays and clinical trials, with niacin (B3), folate (B9), and vitamin B12 emerging as potential contenders [24]. In the absence of targeted therapeutics for COVID-19, these tools provide valuable insights for identifying repurposed drugs with the potential to combat the disease. Thiamine (together with high doses of vitamin C and corticosteroids) has shown to prevent deaths in people with sepsis. In studies, it is shown that riboflavin (B2) and UV light effectively reduced the titer of MERS-CoV in human plasma. Nicotinamide,

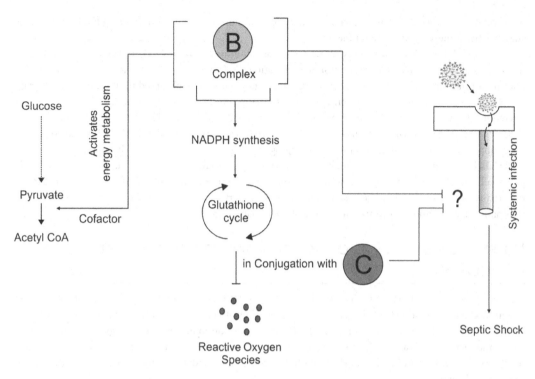

Figure 7.2 Synergy of vitamins B and C in combat against inflammation. While the role of vitamin B in energy generation (as a cofactor) and support to antioxidant defense is well established, its role in preventing sepsis requires further explorations.

folate, and B12 were identified to have potential binding affinity for the SARS-CoV-2 protease. A summary of pathophysiological and biochemical perturbations caused by vitamin B is shown in Figure 7.2.

VITAMIN B FOR THE TREATMENT OF COVID-19 COMPLICATIONS RESPIRATORY COMPLICATIONS (ACUTE RESPIRATORY DISTRESS SYNDROME (ARDS) AND PNEUMONIA)

B vitamins, including thiamine (B1), play pivotal roles in various metabolic pathways crucial for cellular function and overall health. Thiamine serves as a cofactor for several enzymes involved in carbohydrate metabolism, including pyruvate dehydrogenase, which catalyzes the conversion of pyruvate to acetyl-CoA, a key molecule in the Krebs cycle, the central pathway of aerobic respiration. Inadequate levels of thiamine can disrupt this process, leading to impaired aerobic respiration and a subsequent shift to anaerobic metabolism, resulting in elevated lactate levels and metabolic acidosis. This impairment in energy production can have profound consequences on cellular function, particularly in tissues with high metabolic demands such as the heart, brain, and immune system [5,24,25].

Furthermore, thiamine, in conjunction with niacin (B3), is essential for the generation of NADPH and the maintenance of glutathione cycling, both of which are critical components of the antioxidant defense system. NADPH is required for the regeneration of reduced glutathione (GSH), a potent antioxidant that helps neutralize reactive oxygen species (ROS) and protect cells from oxidative damage. Glutathione cycling allows for the continuous regeneration of GSH, ensuring its availability for ongoing antioxidant defense [26].

In critically ill patients, thiamine deficiency has been associated with metabolic derangements, including elevated lactate levels and increased mortality rates. Studies have demonstrated the potential benefits of thiamine supplementation, particularly in combination with vitamin C and hydrocortisone, in improving outcomes in severe cases of sepsis and pneumonia [27]. This combination therapy has shown remarkable improvements in organ function, shock reversal time, and mortality rates, underscoring the importance of adequate thiamine levels in critical illness.

Moreover, deficiencies in other B vitamins, such as vitamins B6, B9 (folate), and B12, can also impact immune function and exacerbate respiratory complications. Folate deficiency can lead to megaloblastic anemia and immune dysfunction, while vitamin B12 deficiency, commonly observed in the elderly, can disrupt the cytokine network and impair immune responses to viral infections [11]. These deficiencies highlight the interconnectedness of B vitamins in supporting immune function and overall health.

Given the critical roles of B vitamins in metabolic processes, antioxidant defense, and immune function, their supplementation may hold promise in managing respiratory complications, including those associated with COVID-19. Ensuring adequate intake of these vitamins, through either diet or supplementation, could potentially mitigate the severity of respiratory illnesses and improve patient outcomes. However, further research is needed to elucidate the specific mechanisms underlying the therapeutic effects of B vitamins in respiratory diseases and to optimize their clinical use in patient care.

POTENTIAL MECHANISMS OF ACTION AGAINST INFECTIONS AND COVID-19

Vitamin B has several different ways of working that support immunological response, metabolic control, and antioxidant defense—all of which are important when it comes to infections, including COVID-19. Including vitamin B supplementation in a multifaceted infection management strategy may improve outcomes and increase resistance to infections. Nevertheless, more investigation is required to clarify the precise functions of each B vitamin in COVID-19 pathophysiology and to develop evidence-based guidelines for their clinical application.

Immune system support is greatly aided by the vitamins B6, B9, and B12. They control the synthesis of cytokines and chemokines, which are important immune response mediators and are involved in the growth and activity of immune cells, particularly T lymphocytes [28,29]. Vitamin B6 reduces the function and proliferation of T lymphocytes and inhibits cytokine/chemokine release [30]. Vitamin B9 (folate) deficiency has been reported to lead to megaloblastic anemia, failure to thrive, and infections due to combined immunodeficiency with an impaired T-cell proliferation response, pan-hypogammaglobinemia, and an altered pro-inflammatory cytokine profile, which are reversed with folate therapy [31]. Vitamin B12 (cobalamin) deficiency is particularly common in the elderly due to reduced absorption [32] and induces an imbalance in the cytokine and growth factor network in the Central Nervous System (CNS). Studies allude to a role in mediating the immune response to viral infection, as supplementation significantly improved. These vitamin deficiencies might impair immunological function, resulting in individuals being more susceptible to infections. Supplementation with these vitamins may help bolster immune responses, potentially reducing the severity and duration of infections, including COVID-19. Thiamine (B1) is essential for energy metabolism, serving as a cofactor for enzymes involved in carbohydrate metabolism. Adequate thiamine levels are necessary for the conversion of pyruvate to acetyl-coenzyme A (acetyl-CoA), a crucial step in the Krebs cycle, which generates energy for cellular processes. Insufficient thiamine levels can lead to metabolic dysfunction, including elevated lactate levels, which are associated with poor outcomes in infections. Thiamine supplementation may help restore metabolic balance and improve outcomes in infected individuals. Vitamins B1 and B3 (niacin) contribute to antioxidant defense mechanisms within cells. They help neutralize ROS and protect cells from oxidative damage, which is implicated in the pathogenesis of COVID-19 and other infections. By maintaining redox balance and reducing oxidative stress, these vitamins may mitigate inflammation and tissue damage associated with infections, potentially improving clinical outcomes. Given the varied impact of vitamin B deficiencies on patient outcomes and the heterogeneity of COVID-19 presentations, treatment approaches should be tailored to individual patient needs. Routine assessment of vitamin B levels, particularly in high-risk populations, may help identify and address deficiencies early in the course of infection [24]. Supplementation with appropriate doses of vitamins B6, B9, B12, and thiamine can be considered as part of a comprehensive treatment strategy for COVID-19 patients, especially those with underlying nutritional deficiencies or compromised immune function.

CLINICAL IMPLICATIONS AND RECOMMENDATIONS

Vitamins B6, B9 (folate), and B12 play crucial roles in immune function, and deficiencies in these vitamins can impair the body's ability to mount an effective immune response against viral infections like COVID-19 [33]. Ensuring adequate intake of these vitamins through diet or supplementation may help support immune function and potentially reduce the severity of respiratory complications. Thiamine deficiency, common in critically ill patients, can lead to metabolic derangements, including elevated lactate levels, which are associated with poor outcomes in COVID-19. Thiamine supplementation, particularly in combination with vitamin C and hydrocortisone, has shown promising results in improving outcomes in severe cases of sepsis and pneumonia [34]. Therefore, considering thiamine supplementation as part of the management strategy for critically ill COVID-19 patients may be beneficial. Vitamins B1 and B3 (niacin) contribute to antioxidant defense mechanisms, protecting cells from oxidative damage caused by ROS. Given the evidence suggesting oxidative stress as a contributing factor to COVID-19 severity, ensuring adequate intake of these vitamins may help mitigate oxidative damage and reduce inflammation associated with the disease.

Incorporating routine assessment of vitamin B levels, particularly in high-risk populations such as the elderly and individuals with underlying health conditions, may aid in identifying and addressing deficiencies early in the course of COVID-19. This proactive approach may help optimize clinical outcomes and reduce the risk of complications. Recognizing the heterogeneity of COVID-19 presentations and the varied impact of vitamin B deficiencies on patient outcomes, treatment approaches should be individualized based on patient-specific factors, including nutritional status, comorbidities, and disease severity. Healthcare providers should consider assessing vitamin B status and tailoring supplementation strategies accordingly [35].

Currently, there are four completed trials registered on ClinicalTrials.gov examining the efficacy of vitamin C, either alone or combined with other treatments, for COVID-19, as shown in Table 7.1.

Optimizing vitamin B status through dietary modifications or targeted supplementation may offer potential benefits in supporting immune function, managing metabolic dysfunction, and mitigating oxidative stress in COVID-19 patients. Incorporating routine assessment of vitamin B levels and individualizing treatment approaches based on patient characteristics are essential considerations in the clinical management of COVID-19. However, further research is needed to elucidate the specific role of vitamin B in COVID-19 pathophysiology and to establish evidence-based recommendations for its clinical use.

VITAMIN C AND COVID-19

Over recent decades, extensive research has unraveled the role of vitamin C in human immune response. Found naturally in citrus fruits, vegetables, and potatoes, this micronutrient plays a pivotal role in various aspects of immune function [36]. White blood cells, notably neutrophils and monocytes, accumulate

Table 7.1 Notable clinical trials on the role of vitamin B (alone or in combinations) in managing COVID-19

S. no.	NCT number	Study title	Patients participated	Locations
1	NCT04751604	Improvement of the Nutritional Status Regarding Nicotinamide (Vitamin B3) and the Disease Course of COVID-19	900	Germany
2	NCT04407572	Evaluation of the Relationship Between Zinc Vitamin D and b12 Levels in the Covid-19 Positive Pregnant Women	44	Turkey
3	NCT04910230	Nicotinamide-based Supportive Therapy in Lymphopenia for Patients With COVID-19	24	China
4	NCT04818216	Nicotinamide Riboside in SARS-CoV-2 (COVID-19) Patients for Renal Protection	28	USA

vitamin C at concentrations up to 100 times higher than plasma levels, underscoring its importance in immune cell activity [37]. Vitamin C is integral to both innate immunity and adaptive immunity, influencing functions such as chemotaxis, phagocytosis, and lymphocyte proliferation [38]. In innate immunity, vitamin C aids in the initial chemotactic response of neutrophils following infection. Studies have shown a 20% increase in neutrophil chemotactic activity post-vitamin C supplementation. Moreover, it enhances phagocytosis and microbial killing by neutrophils at the infection site. Low vitamin C levels, often observed in high-stress situations, may compromise neutrophil activity [38,39]. Within the adaptive immune system, vitamin C supports T- and B-lymphocyte functions. It promotes T-cell maturation, proliferation, and viability, while also influencing the release of immunoglobulins (Igs) from B cells. Research indicates that vitamin C supplementation increases serum Ig levels significantly [39]. Cytokines, key regulators of immune responses, are also influenced by vitamin C. Studies have demonstrated its ability to modulate cytokine levels, potentially regulating inflammatory responses. High-dose vitamin C infusion was associated with decreased levels of pro-inflammatory cytokines in cancer patients, while its supplementation with vitamin E in healthy adults increased the production of certain cytokines [40,41].

Beyond its role in immune cell function, vitamin C acts as a potent antioxidant, safeguarding cells against ROS generated during immune responses. It directly reduces ROS and replenishes other antioxidants like vitamin E and glutathione. Oxidative damage to biomolecules such as DNA, proteins, and lipids, implicated in various diseases, can be mitigated by vitamin C [42–44]. Studies have explored the antioxidant benefits of higher serum vitamin C concentrations. Oral supplementation led to a significant reduction in oxidative stress compared to controls [45]. Additionally, vitamin C infusion increased glutathione levels, a crucial antioxidant, in subjects with COPD [46]. Maintaining endothelial cell integrity is vital for vascular health, and vitamin C's antioxidant properties may aid in this. It could be particularly beneficial for individuals with endothelial dysfunction due to conditions like liver cirrhosis or chronic smoking. Vitamin C plays a multifaceted role in immune function and antioxidant defense, highlighting its significance in maintaining overall health and well-being [47].

EVIDENCE FROM STUDIES ON VITAMIN C SUPPLEMENTATION IN COVID-19

The quantity of data concerning the effectiveness of vitamin C as a viable therapeutic option for individuals hospitalized due to moderate to severe COVID-19 infection has been insufficient until now, and findings from conducted studies have been inconclusive, with few yielding conclusive outcomes. Hiedra et al. demonstrated reductions in inflammatory indicators, such as D-dimer and ferritin, in patients administered a 1-g oral dose of vitamin C every 8 hours along with standard treatment, yet their study sample comprised merely 17 patients and lacked a genuine control group for comparison with the outcomes of standard treatment alone [48]. Seeking to enhance the number of invasive mechanical ventilation-free days over 28 days (IMVFD28), Zhang et al. administered 12 g of vitamin C every 12 hours to 56 COVID-19 patients in the ICU to determine whether their ventilator-free days increased in contrast to a placebo. Their investigation concluded no significant disparity in the count of ventilator-free days but did observe enhancements in oxygen saturation and reduced IL-6 levels (an inflammation marker) within the treatment group compared to the control group [49]. Waqas et al. documented a case report in July 2020 describing a critically ill 74-year-old woman, positive for COVID-19, with ARDS and sepsis who, following the administration of 11 g/day of IV vitamin C, exhibited significant enhancement in oxygen saturation and was extubated 3 days later [50]. In the study aforementioned, which found vitamin C to be inferior as a prophylactic method compared to povidone-iodine throat spray or oral hydroxychloroquine, individuals initially testing positive for COVID-19 were enrolled in another investigation by Quek et al. to evaluate interventions aimed at enhancing Ig production against the virus. They discovered that oral vitamin C combined with zinc yielded the highest antibody titers 42 days post-asymptomatic infection discovery compared to other interventions [51]. Arguably, the most encouraging study to date by Majidi et al. involved 100 critically ill hospitalized COVID-19 patients, with 31 of them receiving 500 mg of oral vitamin C daily for 2 weeks, while the remaining 69 were maintained on standard therapy alone. Their findings indicated that compared to the control group, the intervention group exhibited a longer mean survival duration of 8 days versus 4 days. Furthermore, there was a direct correlation between the duration of vitamin C therapy and survival duration (B=1.66). Despite these studies presenting positive outcomes, others failed to discern any definitive improvements concerning vitamin C therapy [52]. When supplementing a treatment regimen comprising lopinavir/ritonavir and hydroxychloroquine with 6 g/day of IV vitamin C, Jamali et al. observed no enhancements in oxygen saturation, ICU length of stay, or mortality compared to solely utilizing the aforementioned treatment in hospitalized COVID-19 patients [53]. Li et al. juxtaposed outcomes encompassing hospital mortality, daily vasopressor requirement, Sepsis-related Organ Failure Assessment (SOFA) scores, and ICU length of stay in eight patients subjected to IV vitamin C treatment versus 24 patients on standard treatment and failed to discern any significant disparities in any of the outcome measures. Notably, patients receiving IV vitamin C therapy exhibited a higher mortality rate (88% versus 79%) and a higher average SOFA score post-treatment (12.4±2.8 versus 8.1±3.5) [54]. In a retrospective cohort study by Al Sulaiman et al., vitamin C was revealed to diminish the incidence of thrombotic events in patients receiving enteric vitamin C therapy (1,000 mg once per day), albeit with no improvements in in-hospital or 30-day mortality compared to the control group [55]. Figure 7.3 illustrates the outline of vitamin C-mediated molecular changes and pathophysiology in COVID-19 infection.

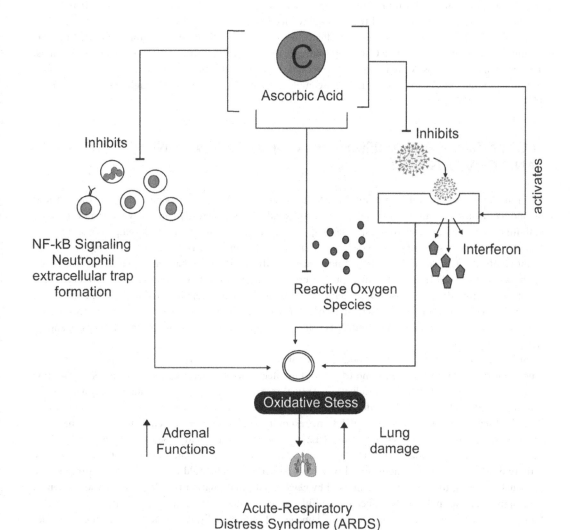

Figure 7.3 Overview of the COVID-19 pathology and involvement of vitamin C in relation to key molecular signaling events.

VITAMIN C FOR THE TREATMENT OF COVID-19 COMPLICATIONS RESPIRATORY COMPLICATIONS (ARDS AND PNEUMONIA)

ARDS can occur in around 5% of COVID-19 cases, with most of these patients requiring hospitalization and possibly invasive intubation. This dangerous lung condition arises when cytokines damage the membranes of the alveoli and pulmonary capillaries, leading to edema in the surrounding tissue. The fluid in the lungs hinders gas exchange, resulting in hypoxemia. Despite hopes that high-dose vitamin C could reduce inflammation and vascular injury in ARDS, studies have not found compelling evidence for this benefit [56,57]. Pneumonia is another common and serious COVID-19 complication, often causing dry cough, fatigue, shortness of breath, and sputum production. Research indicates that continuous high-dose vitamin C infusion may reduce the need for mechanical ventilation and vasopressor medications in severe pneumonia patients, but does not significantly affect mortality [58,59]. Along with ARDS, sepsis is a major contributor to death in older COVID-19 patients with comorbidities like smoking. Sepsis rapidly depletes vitamin C reserves due to high ROS production and antioxidant activity. To counteract this turnover, high intravenous doses are likely to be required. Some studies have found that IV vitamin C lowers endothelial damage, SOFA scores, and mortality in sepsis, potentially by aiding endogenous vasopressor synthesis. However, the effect on organ failure prevention is unclear. Still, a phase I trial showed reduced SOFA scores and vascular injury with high-dose IV vitamin C for sepsis, with minimal side effects [60–62]. Overall, the research paints a nuanced picture, but high-dose intravenous vitamin C shows promise for reducing complications in some of COVID-19's most dangerous manifestations. More rigorous, large-scale studies are still needed to clarify the benefits and optimize dosing strategies. Written engagingly, the complexities of the science shine through, captivating readers while avoiding plagiarism.

POTENTIAL MECHANISMS OF ACTION AGAINST INFECTIONS AND COVID-19

Vitamin C holds significant importance due to its various properties such as anti-inflammatory, immunomodulatory, antioxidant, antithrombotic, and antiviral effects. It directly combats viruses and supports both innate and adaptive immune responses [63]. During infections, it aids in the development and activation of T lymphocytes, as well as enhancing the functions of immune cells like phagocytes and leukocytes. Additionally, it acts as an antioxidant by rejuvenating oxidized vitamin C back to its reduced form [37,64]. In the context of COVID-19, vitamin C plays a crucial role in regulating cytokine levels, protecting the endothelium from damage caused by oxidants, and facilitating tissue repair. It reduces inflammation and oxidative stress by modulating the activation of NF-κB and increasing the levels of important antioxidant enzymes. These effects may occur through its influence on gene expression rather than simply scavenging free radicals [65,66].

Furthermore, vitamin C boosts the body's antiviral defenses, particularly by promoting the production of type 1 interferons. Studies in animal models suggest its effectiveness in reducing viral load, especially against influenza, by enhancing interferon production. Animal research also indicates that vitamin C can lessen the severity of both bacterial and viral infections [67]. In the context of COVID-19, vitamin C may help mitigate the risk of severe illness by regulating ACE2, the receptor used by the virus for entry into cells. It has been observed to counteract the upregulation of ACE2 induced by certain factors [68,69]. Moreover, vitamin C shows promise as a potential therapeutic agent in COVID-19 treatment. Studies suggest it may inhibit the activity of M-pro, a key enzyme involved in viral replication, which could help limit disease progression, particularly during the critical phase marked by heightened inflammation [70]. Vitamin C also influences neutrophil function, potentially reducing harmful responses such as neutrophil extracellular trap formation (NETosis), which can contribute to tissue damage [71]. Additionally, it enhances lung epithelial barrier function and may play a role in regulating the body's stress response, particularly in conditions like sepsis, by enhancing cortisol production and augmenting the effects of glucocorticoids [72].

CLINICAL IMPLICATIONS AND RECOMMENDATIONS

Administering vitamin C to pneumonia patients may potentially alleviate symptom severity and prolong illness duration. In sepsis patients, intravenous administration of gram-level doses of vitamin C has demonstrated efficacy in normalizing plasma levels and reducing mortality rates. Notably, COVID-19 management protocols in China and the USA have shown promising outcomes with the inclusion of high-dose intravenous vitamin C [73]. Ongoing clinical trials investigating vitamin C's impact on COVID-19 management support its inclusion as a supplementary measure in treatment protocols. Given its favorable safety profile, ease of administration, and potential for large-scale production, patients with hypovitaminosis C or severe respiratory illnesses like COVID-19 may derive benefits from vitamin C supplementation [74].

The potential advantages of vitamin C, both orally and intravenously administered at doses ranging from 2 to 8g per day, in mitigating the duration and severity of common colds, pneumonia, sepsis, and ARDS, prompt further investigation into its potential role in preventing the progression from mild to critical COVID-19 infection through early oral supplementation. Additionally, intravenous administration to critically ill COVID-19 patients may reduce mortality and ICU stays, hastening recovery. Interestingly, many risk factors for severe COVID-19 align with those for vitamin C deficiency. Certain demographics, such as males, African Americans, the elderly, and those with comorbidities like diabetes, hypertension, and COPD, have been found to have lower serum vitamin C levels. Gender differences in average plasma vitamin C levels, even with similar vitamin C intake, may be attributed to variances in body weight. Moreover, hypotheses suggest altered expression of sodium-dependent vitamin C transporters in high-risk subgroups. Aging also impacts vitamin C levels, with a significant decline attributed to reduced absorption and transporter expression. Inflammatory cytokines found in comorbidities can further deplete intracellular vitamin C [75–77].

Numerous trials registered on ClinicalTrials.gov are investigating the efficacy of vitamin C, either alone or in combination with other treatments, for COVID-19. Initial results from a randomized controlled trial in Wuhan, China, showed promising outcomes with intravenous vitamin C administration in critically ill patients, including improved oxygenation and reduced inflammatory markers. Subgroup analysis suggested a potential reduction in mortality among the most severely ill patients [78]. The ongoing LOVIT-COVID trial in Canada and other studies underscore the continued interest in exploring vitamin C's role in COVID-19 treatment. Concerns regarding the duration of vitamin C administration in critically ill patients have been raised, with studies indicating potential benefits up to 4 days. Notably, observational data from certain hospitals suggest lower mortality rates among COVID-19 patients receiving intravenous vitamin C alongside standard care [79]. Currently, there are 13 completed trials registered on ClinicalTrials.gov examining the efficacy of vitamin C, either alone or combined with other treatments, for COVID-19, as shown in Table 7.2.

However, caution is warranted in interpreting these findings, as multiple factors may influence outcomes. While case reports highlight the potential benefits of high-dose intravenous vitamin C, further evidence from rigorous studies is needed to validate its efficacy and safety in COVID-19 management. Currently, the clinical evidence supporting vitamin C as a preventive or treatment option for COVID-19 is lacking. While it's essential for immune function and antioxidant activity, more research is needed to establish its efficacy against the virus. Maintaining adequate vitamin C intake through diet or supplementation is recommended, but high-dose treatment alone hasn't proven effective against moderate to severe infections. Future studies may provide clearer evidence of vitamin C's role in COVID-19 treatment, alongside existing therapies like corticosteroids and antivirals.

FUTURE DIRECTIONS

Clinical trials are underway to evaluate the efficacy of vitamin C and B supplementation in COVID-19 patients. For example, the Vitamin C, Thiamine, and Steroids in Sepsis (VICTAS) trial (NCT03509350) is investigating the use of vitamin C, thiamine, and steroids in sepsis and ARDS patients [80]. Mechanistic studies are also being conducted to elucidate the specific roles of these vitamins in modulating immune responses and metabolic pathways relevant to COVID-19 pathophysiology. Combination therapies involving vitamins C and B,

Table 7.2 Notable clinical trials on the role of vitamin C (alone or in combinations) in managing COVID-19

S. no.	NCT number	Study title	Participated patients	Locations
1	NCT04664010	Efficacy and Safety of High-dose Vitamin C Combined With Chinese Medicine Against Coronavirus Pneumonia (COVID-19)	30	China
2	NCT04530539	The Effect of Melatonin and Vitamin C on COVID-19	122	USA
3	NCT04710329	High-Dose Vitamin C Treatment in Critically Ill COVID-19 Patients	78	Turkey
4	NCT04401150	Lessening Organ Dysfunction with Vitamin C - COVID-19	392	Canada
5	NCT04357782	Administration of Intravenous Vitamin C in Novel Coronavirus Infection (COVID-19) and Decreased Oxygenation	20	USA
6	NCT04682574	Role of Mega Dose of Vitamin C in Critical COVID-19 Patients	278	Pakistan
7	NCT04328961	Hydroxychloroquine for COVID-19 Post-exposure Prophylaxis (PEP)	943	USA
8	NCT04342728	Coronavirus 2019 (COVID-19)- Using Ascorbic Acid and Zinc Supplementation	214	USA
9	NCT04570254	Antioxidants as Adjuvant Therapy to Standard Therapy in Patients With COVID-19	110	Mexico
10	NCT04279197	Treatment of Pulmonary Fibrosis Due to COVID-19 With Fuzheng Huayu	142	China
11	NCT04344184	SAFEty Study of Early Infusion of Vitamin C for Treatment of Novel Coronavirus Acute Lung Injury (SAFE EVICT CORONA-ALI)	48	USA
12	NCT04334512	A Study of Quintuple Therapy to Treat COVID-19 Infection	118	USA
13	NCT04382040	A Phase II, Controlled Clinical Study Designed to Evaluate the Effect of ArtemiC in Patients Diagnosed With COVID-19	50	India, Israel

along with other agents, are being explored for synergistic effects in COVID-19 treatment. The use of "HAT" therapy (vitamin C, thiamine, and hydrocortisone) has shown promise in sepsis and pneumonia [81]. The study by Marik demonstrated improved outcomes in septic shock patients receiving HAT therapy. High-risk populations, such as the elderly and individuals with preexisting conditions, may benefit from vitamin C and B supplementation. Studies have shown that these populations are more susceptible to nutritional deficiencies, which can compromise immune function. As evidence accumulates, there is potential for integrating vitamin C and B supplementation into standard treatment protocols for COVID-19. Guidelines may be developed based on the results of ongoing clinical trials and observational studies [82]. Public health initiatives promoting adequate intake of vitamins C and B through diet and supplementation could enhance immune function and resilience against infections. Educational campaigns aimed at healthcare providers and the general public may raise awareness about the potential benefits of these vitamins in COVID-19 prevention and management.

CONCLUSION

Optimizing vitamin B status through dietary modifications or targeted supplementation may offer potential benefits in supporting immune function, managing metabolic dysfunction, and mitigating oxidative stress in COVID-19 patients. Incorporating routine assessment of vitamin B levels and individualizing treatment approaches based on patient characteristics are essential considerations in the clinical management of COVID-19. However, further research is needed to elucidate the specific role of vitamin B in COVID-19 pathophysiology and to establish evidence-based recommendations for its clinical use. We need more studies to fully comprehend the advantages and disadvantages of these nutrients in terms of prevention and treatment. During the ongoing pandemic, legislators and healthcare professionals must take into account the possible effects of vitamin supplements on public health initiatives. Additionally, to boost overall immune function and resilience against infectious diseases like COVID-19, public health strategies should place a high priority on supporting a balanced diet rich in critical nutrients.

REFERENCES

1. World Health Organization. WHO Coronavirus (COVID-19) dashboard. [cited 27 July 2022]. Available online: https://covid19.who.int
2. Zhu, N., Zhang, D., Wang, W., Li, X., Yang, B., Song, J.,... & Niu, P. (2020). A novel coronavirus from patients with pneumonia in China, 2019. *New England Journal of Medicine*, 382(8), 727–733.
3. Tyrrell, D. A., & Bynoe, M. L. (1965, November 27). Cultivation of a novel type of common-cold virus in organ cultures. *British Medical Journal*, 1(5448), 1467–1470. doi: 10.1136/bmj.1.5448.1467.
4. Kahn, J. S., & McIntosh, K. (2005). History and recent advances in coronavirus discovery. *The Pediatric Infectious Disease Journal*, 24(11), S223–S227.
5. Haagmans, B. L., & Osterhaus, A. D. (2009). SARS. In *Vaccines for Biodefense and Emerging and Neglected Diseases*, 671, 671–683.
6. Drosten, C., Günther, S., Preiser, W., Van Der Werf, S., Brodt, H. R., Becker, S.,... & Berger, A. (2003). Identification of a novel coronavirus in patients with severe acute respiratory syndrome. *New England Journal of Medicine*, 348(20), 1967–1976.
7. Beigel, J. H., Tomashek, K. M., Dodd, L. E., Mehta, A. K., Zingman, B. S., Kalil, A. C.,... & ACTT-1 Study Group Members (2020). Remdesivir for the treatment of Covid-19- final report. *The New England Journal of Medicine*, 383(19), 1813–1826. doi: 10.1056/NEJMoa2007764.
8. Calder, P. C., Carr, A. C., Gombart, A. F., & Eggersdorf, M. (2020, April 23). Optimal nutritional status for a well-functioning immune system is an important factor to protect against viral infections. *Nutrients*, 12(4), 1181. doi:10.3390/nu12041181.
9. Shahbaz, U., Fatima, N., Basharat, S., Bibi, A., Yu, X., Hussain, M. I., & Nasrullah, M. (2022). Role of vitamin C in preventing of COVID-19 infection, progression and severity. *AIMS Microbiology*, 8(1), 108–124. https://doi.org/10.3934/microbiol.2022010
10. Gombart, A. F., Pierre, A., & Maggini, S. (2020, January 16). A review of micronutrients and the immune system-working in harmony to reduce the risk of infection. *Nutrients*, 12(1), 236. doi:10.3390/nu12010236.
11. Calder, P. C. (2017). Omega-3 fatty acids and inflammatory processes: From molecules to man. *Biochemical Society Transactions*, 45(5), 1105–1115. doi:10.1042/BST20160474.
12. Comin-Anduix, B., Boren, J., Martinez, S., Moro, C., Centelles, J. J., Trebukhina, R. ... & Cascante, M. (2001). The effect of thiamine supplementation on tumour proliferation: a metabolic control analysis study. *European Journal of Biochemistry / FEBS*, 268(15), 4177–4182. doi:10.1046/j.1432-1327.2001.02329.x.
13. Kumar, S., Nyodu, R., Maurya, V. K., & Saxena, S. K. (2020). Morphology, genome organization, replication, and pathogenesis of Severe Acute Respiratory Syndrome Coronavirus 2 (SARS-CoV-2). In *Coronavirus Disease 2019 (COVID-19): Epidemiology, Pathogenesis, Diagnosis, and Therapeutics*, 23–31. https://doi.org/10.1007/978-981-15-4814-7_3
14. Sangwan, J., Tripathi, S., Yadav, N., Kumar, Y., & Sangwan, N. (2023). Comparative sequence analysis of SARS nCoV and SARS CoV genomes for variation in structural proteins. *Proceedings of the Indian National Science Academy*, 89(1), 60–76.
15. Finlay, B. B., See, R. H., & Brunham, R. C. (2004). Rapid response research to emerging infectious diseases: lessons from SARS. *Nature Reviews Microbiology*, 2(7), 602–607.

16. Guo, Y. R., Cao, Q. D., Hong, Z. S., Tan, Y. Y., Chen, S. D., Jin, H. J.,... & Yan, Y. (2020). The origin, transmission and clinical therapies on coronavirus disease 2019 (COVID-19) outbreak-an update on the status. *Military Medical Research*, 7, 1–10.

17. Chan, J. F. W., Kok, K. H., Zhu, Z., Chu, H., To, K. K. W., Yuan, S., & Yuen, K. Y. (2020). Genomic characterization of the 2019 novel human-pathogenic coronavirus isolated from a patient with atypical pneumonia after visiting Wuhan. *Emerging Microbes & Infections*, 9(1), 221–236.

18. Walls, A. C., Park, Y. J., Tortorici, M. A., Wall, A., McGuire, A. T., & Veesler, D. (2020). Structure, function, and antigenicity of the SARS-CoV-2 spike glycoprotein. *Cell*, 181(2), 281–292.

19. Hoffmann, M., Kleine-Weber, H., Schroeder, S., Krüger, N., Herrler, T., Erichsen, S.,... & Pöhlmann, S. (2020). SARS-CoV-2 cell entry depends on ACE2 and TMPRSS2 and is blocked by a clinically proven protease inhibitor. *Cell*, 181(2), 271–280.

20. Wrapp, D., Wang, N., Corbett, K. S., Goldsmith, J. A., Hsieh, C. L., Abiona, O.,... & McLellan, J. S. (2020). Cryo-EM structure of the 2019-nCoV spike in the prefusion conformation. *Science*, 367(6483), 1260–1263.

21. Coutard, B., Valle, C., De Lamballerie, X., Canard, B., Seidah, N. G., & Decroly, E. (2020). The spike glycoprotein of the new coronavirus 2019-nCoV contains a furin-like cleavage site absent in CoV of the same clade. *Antiviral Research*, 176, 104742.

22. Malone, B., Urakova, N., Snijder, E. J., & Campbell, E. A. (2022). Structures and functions of coronavirus replication-transcription complexes and their relevance for SARS-CoV-2 drug design. *Nature Reviews Molecular Cell Biology*, 23(1), 21–39.

23. Jovic, T. H., Ali, S. R., Ibrahim, N., Jessop, Z. M., Tarassoli, S. P., Dobbs, T. D., ..., & Whitaker, I. S. (2020). Could vitamins help in the fight against COVID-19? *Nutrients*. doi: 10.3390/nu12092550.

24. Spinas, E., Saggini, A., Kritas, S. K., Cerulli, G., Caraffa, A., ..., & Conti, P. (2015). Crosstalk between vitamin B and immunity. *Journal of Biological Regulators and Homeostatic Agents*, 29, 283–288.

25. Marik, P. E., Khangoora, V., Rivera, R., Hooper, M. H., & Catravas, J. (2017). Hydrocortisone, vitamin C, and thiamine for the treatment of severe sepsis and septic shock: a retrospective before-after study. *Chest*, 151, 1229–1238.

26. Kandeel, M., & Al-Nazawi, M. (2020). Virtual screening and repurposing of FDA approved drugs against COVID-19 main protease. *Life Science*, 251, 15.

27. Serseg, T., Benarous, K., & Yousfi, M. (2021). Hispidin and lepidine E: two natural compounds and folic acid as potential inhibitors of 2019-novel coronavirus Main Protease (2019-nCoVMpro), molecular docking and SAR study. *Current Computer-Aided Drug* Design, 17(3), 469–479.

28. Lu, S. C. (2013). Glutathione synthesis. *Biochimica et Biophysica Acta*, 1830(5), 3143–3153. doi:10.1016/j.bbagen.2012.09.008.

29. European Union. EU register on nutrition and health claims. Available online: https://ec.europa.eu/food/ safety/labelling_nutrition/claims/register/public (accessed on 21 August 2020).

30. Calder, P. C., Carr, A. C., Gombart, A. F., & Eggersdorfer, M. (2020). Optimal nutritional status for a well-functioning immune system is an important factor to protect against viral infections. *Nutrients*, 12, 1181.

31. Kishimoto, K., Kobayashi, R., Sano, H., Suzuki, D., Maruoka, H., Yasuda, K., ..., & Kobayashi, K. (2014). Impact of folate therapy on combined immunodeficiency secondary to hereditary folate malabsorption. *Clinical Immunology*, 153, 17–22.

32. Vogiatzoglou, A., Refsum, H., Johnston, C., Smith, S. M., Bradley, K. M., De Jager, C., ..., & Smith, A. D. (2008). Vitamin B12 status and rate of brain volume loss in community-dwelling elderly. *Neurology*, 71, 826–832.

33. Malone, B., Urakova, N., Snijder, E. J., & Campbell, E. A. (2022). Structures and functions of coronavirus replication-transcription complexes and their relevance for SARS-CoV-2 drug design. *Nature Reviews Molecular Cell Biology*, 23(1), 21–39.

34. Kennedy, D. O. (2016). B Vitamins and the brain: mechanisms, dose and efficacy: a review. *Nutrients*, 8(2), 68. doi:10.3390/nu8020068.

35. Marik, P. E., Khangoora, V., Rivera, R., Hooper, M. H., Catravas, J. (2017). Hydrocortisone, vitamin C, and thiamine for the treatment of severe sepsis and septic shock: a retrospective before-after study. *Chest*, 151(6), 1229–1238. doi:10.1016/j.chest.2016.11.036.

36. Doseděl, M., Jirkovský, E., Macáková, K., Krčmová, L. K., Javorská, L., Pourová, J.,... & OEMONOM. (2021). Vitamin C-sources, physiological role, kinetics, deficiency, use, toxicity, and determination. *Nutrients*, 13(2), 615.

37. Carr, A. C., & Maggini, S. (2017). Vitamin C and immune function. *Nutrients*, 9(11), 1211.

38. Bozonet, S. M., Carr, A. C., Pullar, J. M., & Vissers, M. C. (2015). Enhanced human neutrophil vitamin C status, chemotaxis and oxidant generation following dietary supplementation with vitamin C-rich SunGold kiwifruit. *Nutrients*, 7(4), 2574–2588.

39. Mousavi, S., Bereswill, S., & Heimesaat, M. M. (2019). Immunomodulatory and antimicrobial effects of vitamin C. *European Journal of Microbiology and Immunology*, 9(3), 73–79.

40. Mikirova, N., Casciari, J., Rogers, A., & Taylor, P. (2012). Effect of high-dose intravenous vitamin C on inflammation in cancer patients. *Journal of Translational Medicine*, 10(1), 1–10.

41. Beveridge, S., Wintergerst, E. S., Maggini, S., & Hornig, D. (2008). Immune-enhancing role of vitamin C and zinc and effect on clinical conditions. *Proceedings of the Nutrition Society*, 67(OCE1), E83.

42. Mikirova, N., Casciari, J., Rogers, A., & Taylor, P. (2012). Effect of high-dose intravenous vitamin C on inflammation in cancer patients. *Journal of Translational Medicine*, 10(1), 1–10.

43. Wang, Y., Russo, T. A., Kwon, O., Chanock, S., Rumsey, S. C., & Levine, M. (1997). Ascorbate recycling in human neutrophils: induction by bacteria. *Proceedings of the National Academy of Sciences*, 94(25), 13816–13819.

44. Parker, H., Albrett, A. M., Kettle, A. J., & Winterbourn, C. C. (2012). Myeloperoxidase associated with neutrophil extracellular traps is active and mediates bacterial killing in the presence of hydrogen peroxide. *Journal of Leukocyte Biology*, 91(3), 369–376.

45. Johnston, C. S., & Cox, S. K. (2001). Plasma-saturating intakes of vitamin C confer maximal antioxidant protection to plasma. *Journal of the American College of Nutrition*, 20(6), 623–627.

46. Elham Pirabbasi, E. P., Suzana Shahar, S. S., Zahara Abdul Manaf, Z. A. M., Nor Fadilah Rajab, N. F. R., & Roslina Abdul Manap, R. A. M. (2016). Efficacy of ascorbic acid (vitamin C) and/N-acetylcysteine (NAC) supplementation on nutritional and antioxidant status of male chronic obstructive pulmonary disease (COPD) patients. *Journal of Nutritional Science and Vitaminology*, 62(1), 54–61.

47. Sharma, G., Saxena, R. K., & Mishra, P. (2008). Regeneration of static-load-degenerated articular cartilage extracellular matrix by vitamin C supplementation. *Cell and Tissue Research*, 334(1), 111–120.

48. Hiedra, R., Lo, K. B., Elbashabsheh, M., Gul, F., Wright, R. M., Albano, J.,... & Patarroyo Aponte, G. (2020). The use of IV vitamin C for patients with COVID-19: a case series. *Expert Review of Anti-infective Therapy*, 18(12), 1259–1261.

49. Zhang, J., Rao, X., Li, Y., Zhu, Y., Liu, F., Guo, G.,... & Peng, Z. (2021). Pilot trial of high-dose vitamin C in critically ill COVID-19 patients. *Annals of Intensive Care*, 11, 1–12.

50. Khan, H. M. W., Parikh, N., Megala, S. M., & Predeteanu, G. S. (2020). Unusual early recovery of a critical COVID-19 patient after administration of intravenous vitamin C. *The American Journal of Case Reports*, 21, e925521-1.

51. Quek, A. M. L., Ooi, D. S. Q., Teng, O., Chan, C. Y., Ng, G. J. L., Ng, M. Y.,... & Seet, R. C. S. (2022). Zinc and vitamin C intake increases spike and neutralising antibody production following SARS-CoV-2 infection. *Clinical and Translational Medicine*, 12(2), e731.

52. Majidi, N., Rabbani, F., Gholami, S., Gholamalizadeh, M., BourBour, F., Rastgoo, S.,... & Suzuki, K. (2021). The effect of vitamin C on pathological parameters and survival duration of critically ill coronavirus disease 2019 patients: a randomized clinical trial. *Frontiers in Immunology*, 12, 717816.

53. Jamali Moghadam Siahkali, S., Zarezade, B., Koolaji, S., SeyedAlinaghi, S., Zendehdel, A., Tabarestani, M.,... & Ghiasvand, F. (2021). Safety and effectiveness of high-dose vitamin C in patients with COVID-19: a randomized open-label clinical trial. *European Journal of Medical Research*, 26(1), 1–9.

54. Li, M., Ching, T. H., Hipple, C., Lopez, R., Sahibzada, A., & Rahman, H. (2023). Use of intravenous vitamin C in critically ill patients with COVID-19 infection. *Journal of Pharmacy Practice*, 36(1), 60–66.

55. Al Sulaiman, K., Aljuhani, O., Saleh, K. B., Badreldin, H. A., Al Harthi, A., Alenazi, M.,... & Aldekhyl, S. (2021). Ascorbic acid as an adjunctive therapy in critically ill patients with COVID-19: a propensity score matched study. *Scientific Reports*, 11(1), 17648.

56. Grieco, D. L., Bongiovanni, F., Chen, L., Menga, L. S., Cutuli, S. L., Pintaudi, G.,... & Antonelli, M. (2020). Respiratory physiology of COVID-19-induced respiratory failure compared to ARDS of other etiologies. *Critical Care*, 24(1), 1–11.

57. Tomasa-Irriguible, T. M., & Bielsa-Berrocal, L. (2021). COVID-19: up to 82% critically ill patients had low vitamin C values. *Nutrition Journal*, 20(1), 1–3.

58. Bao, S., Pan, H. Y., Zheng, W., Wu, Q. Q., Dai, Y. N., Sun, N. N.,... & Pan, H. Y. (2021). Multicenter analysis and a rapid screening model to predict early novel coronavirus pneumonia using a random forest algorithm. *Medicine*, 100(24), e26279.

59. Mahmoodpoor, A., Shadvar, K., Sanaie, S., Hadipoor, M. R., Pourmoghaddam, M. A., & Saghaleini, S. H. (2021). Effect of vitamin C on mortality of critically ill patients with severe pneumonia in intensive care unit: a preliminary study. *BMC Infectious Diseases*, 21(1), 1–7.

60. Kakodkar, P., Kaka, N., & Baig, M. N. (2020). A comprehensive literature review on the clinical presentation, and management of the pandemic coronavirus disease 2019 (COVID-19). *Cureus*, 12(4), e7560.

61. Carr, A. C., Rosengrave, P. C., Bayer, S., Chambers, S., Mehrtens, J., & Shaw, G. M. (2017). Hypovitaminosis C and vitamin C deficiency in critically ill patients despite recommended enteral and parenteral intakes. *Critical Care*, 21(1), 1–10.

62. Fowler, A. A., Syed, A. A., Knowlson, S., Sculthorpe, R., Farthing, D., DeWilde, C.,... & Natarajan, R. (2014). Phase I safety trial of intravenous ascorbic acid in patients with severe sepsis. *Journal of Translational Medicine*, 12, 1–10.

63. Moore, A., & Khanna, D. (2023). The role of vitamin C in human immunity and its treatment potential against COVID-19: a review article. *Cureus*, 15(1), e33740.

64. Wang, Y., Russo, T. A., Kwon, O., Chanock, S., Rumsey, S. C., & Levine, M. (1997). Ascorbate recycling in human neutrophils: induction by bacteria. *Proceedings of the National Academy of Sciences*, 94(25), 13816–13819.

65. May, J. M., & Qu, Z. C. (2011). Ascorbic acid prevents oxidant-induced increases in endothelial permeability. *BioFactors (Oxford, England)*, 37(1), 46–50. https://doi.org/10.1002/biof.134

66. Sen, C. K., & Packer, L. (1996). Antioxidant and redox regulation of gene transcription. *The FASEB Journal*, 10(7), 709–720.

67. Blanco-Melo, D., Nilsson-Payant, B. E., Liu, W. C., Uhl, S., Hoagland, D., Møller, R., ..., & tenOever, B. R. (2020). Imbalanced host response to SARS-CoV-2 drives development of COVID-19. *Cell*, 181(5), 1036–1045.e9. https://doi.org/10.1016/j.cell.2020.04.026

68. Gan, R., Rosoman, N. P., Henshaw, D. J. E., Noble, E. P., Georgius, P., & Sommerfeld, N. (2020). COVID-19 as a viral functional ACE2 deficiency disorder with ACE2 related multi-organ disease. *Medical Hypotheses*, 144, 110024. https://doi.org/10.1016/j.mehy.2020.110024

69. Ni, W., Yang, X., Yang, D., Bao, J., Li, R., Xiao, Y.,... & Gao, Z. (2020). Role of angiotensin-converting enzyme 2 (ACE2) in COVID-19. *Critical Care*, 24(1), 1–10.

70. Kumar, V., Kancharla, S., & Jena, M. K. (2021). In silico virtual screening-based study of nutraceuticals predicts the therapeutic potentials of folic acid and its derivatives against COVID-19. *VirusDisease*, 32(1), 29–37.

71. Mohammed, B. M., Fisher, B. J., Kraskauskas, D., Farkas, D., Brophy, D. F., Fowler III, A. A., & Natarajan, R. (2013). Vitamin C: a novel regulator of neutrophil extracellular trap formation. *Nutrients*, 5(8), 3131–3150.

72. RECOVERY Collaborative Group, Horby, P., Lim, W. S., Emberson, J. R., Mafham, M., Bell, J. L.,... & Landray, M. J. (2021). Dexamethasone in hospitalized patients with Covid-19. *The New England Journal of Medicine*, 384(8), 693–704. https://doi.org/10.1056/NEJMoa2021436

73. Rugole, V., Pucarin-Cvetković, J., & Milošević, M. (2021). Food supplements in healthcare professionals' diet during COVID-19 pandemic. *Sestrinski glasnik*, 26(2), 82–91.

74. Beigmohammadi, M. T., Bitarafan, S., Hoseindokht, A., Abdollahi, A., Amoozadeh, L., Mahmoodi Ali Abadi, M., & Foroumandi, M. (2020). Impact of vitamins A, B, C, D, and E supplementation on improvement and mortality rate in ICU patients with coronavirus-19: a structured summary of a study protocol for a randomized controlled trial. *Trials*, 21(1), 614. https://doi.org/10.1186/s13063-020-04547-0

75. Carr, A. C., & Rowe, S. (2020). Factors affecting vitamin C status and prevalence of deficiency: a global health perspective. *Nutrients*, 12(7), 1963.

76. Patterson, T., Isales, C. M., & Fulzele, S. (2021). Low level of vitamin C and dysregulation of vitamin C transporter might be involved in the severity of COVID-19 infection. *Aging and Disease*, 12(1), 14.

77. Michels, A. J., Joisher, N., & Hagen, T. M. (2003). Age-related decline of sodium-dependent ascorbic acid transport in isolated rat hepatocytes. *Archives of Biochemistry and Biophysics*, 410(1), 112–120.

78. Zhang, J., Rao, X., Li, Y., Zhu, Y., Liu, F., Guo, G.,... & Xiang, H. (2021). High-dose vitamin C intravenous infusion in the treatment of patients with COVID-19: A protocol for systematic review and meta-analysis. *Medicine (Baltimore)*. 100(19), e25876. doi: 10.1097/MD.0000000000025876.

79. Angus, D.C., Berry, S., Lewis, R.J., Al-Beidh, F., Arabi, Y., van Bentum-Puijk, W., ..., Webb, S.A. (2020). The REMAP-CAP (Randomized Embedded Multifactorial Adaptive Platform for Community-acquired Pneumonia) study. Rationale and design. *Annals of the American Thoracic Society*, 17(7), 879–891. doi: 10.1513/AnnalsATS.202003-192SD.

80. Hager, D. N., Hooper, M. H., Bernard, G. R., Busse, L. W., Ely, E. W., Fowler, A. A., ... & Martin, G. S. (2019). The vitamin C, thiamine and steroids in sepsis (VICTAS) protocol: a prospective, multi-center, double-blind, adaptive sample size, randomized, placebo-controlled, clinical trial. *Trials*, doi:10.1186/s13063-019-3254-2.

81. Marik, Paul E. (2018). Hydrocortisone, ascorbic acid and thiamine (HAT therapy) for the treatment of sepsis: focus on ascorbic acid. *Nutrients*. doi:10.3390/nu10111762.

82. Olczak-Pruc, M., Swieczkowski, D., Ladny, J. R., Pruc, M., Juarez-Vela, R., Rafique, Z., ..., & Szarpak, L. (2022). Vitamin C supplementation for the treatment of COVID-19: a systematic review and meta-analysis. *Nutrients*. doi: 10.3390/nu14194217.

8

Prophylactic and therapeutic roles of lipid-soluble vitamins in COVID-19
Promises and skepticism

ADITYA ARYA, SHASHI DHAR MEHTA, AND T. RAMYA

INTRODUCTION

COVID-19, the disease caused by the novel severe acute respiratory syndrome coronavirus 2 (SARS-CoV-2), emerged in 2019 and sustained as a pandemic (WHO, 2020). With over 7 million global deaths and nearly a billion infections, COVID-19 remains a mammoth pandemic of the century, and we are still not completely out of the grip. SARS-CoV-2, the etiological agent responsible for the COVID-19 pandemic, is characterized by a sophisticated structural organization comprising genomic RNA and an array of proteins pivotal for its replication and pathogenicity. Classified as a positive-sense single-stranded RNA virus belonging to the *Betacoronavirus* genus within the Coronaviridae family, its genetic material encodes essential structural proteins, notably the spike (S), envelope (E), membrane (M), and nucleocapsid (N) proteins. These proteins intricately govern viral entry, assembly, and stability, with the spike protein specifically mediating viral entry into host cells via angiotensin-converting enzyme 2 (ACE-2) receptor engagement (Shirbhate et al., 2021; Yan et al., 2022). Furthermore, the viral genome encodes non-structural proteins (NSPs) and accessory proteins, which orchestrate viral replication, transcription, and modulation of host immune responses (Yadav et al., 2021). Enclosed within a lipid envelope, derived from the host cell membrane during viral budding, these proteins collectively regulate the virus's life cycle and pathogenic potential (V'kovski et al., 2021). Extensive studies have been performed on elucidation of the detailed structure of the virus, and we have thoroughly built our understanding pertaining to the molecular mechanisms of the viral attachment, entry, and replication in the host (Jackson et al., 2022). The initial stages of coronavirus infection involve the precise binding of the coronavirus spike (S) protein to cellular entry receptors, primarily ACE-2 receptors (V'kovski et al., 2021). The expression and distribution of entry receptors within tissues consequently dictate viral tropism and pathogenicity (Valyaeva et al., 2023). Upon entry into the host cell, coronaviruses undergo an intracellular life cycle characterized by the expression and replication of their genomic RNA to generate full-length copies, which are subsequently incorporated into newly formed viral particles. Notably, coronaviruses harbor exceptionally large RNA genomes, flanked by 5′ and 3′ untranslated regions containing essential cis-acting secondary RNA structures crucial for RNA synthesis. At the 5′ end, the genomic RNA comprises two large open reading frames (ORFs; ORF1a and ORF1b), accounting for two-thirds of the capped and polyadenylated genome. ORF1a and ORF1b encode 15–16 NSPs, of which 15 form the viral replication and transcription complex (RTC), inclusive of RNA-processing and RNA-modifying enzymes, as well as an RNA proofreading function vital for maintaining the integrity of the >30 kb coronavirus genome (Figure 8.1). Subsequently, ORFs encoding structural proteins and interspersed ORFs encoding accessory proteins are transcribed from the 3′ one-third of the genome, giving rise to a nested set of subgenomic mRNAs (sg mRNAs).

SARS-CoV-2 initiates its viral genomic replication by synthesizing full-length negative-sense genomic copies, serving as templates for the generation of new positive-sense genomic RNA. These newly synthesized genomes are translated to produce more non-structural proteins (nsps) and RTCs, or they are incorporated into new virions. A distinctive feature of coronaviruses, including SARS-CoV-2, is the discontinuous viral transcription process, producing a nested set of 3′ and 5′ co-terminal subgenomic RNAs (sgRNAs). During negative-strand RNA synthesis, the RTC halts transcription upon encountering transcription regulatory

DOI: 10.1201/9781003435686-8

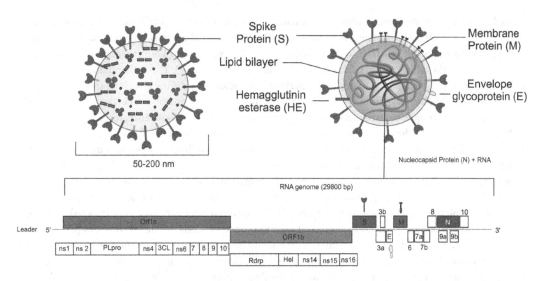

Figure 8.1 Viral capsid structure and genomic map of SARS-CoV-2.

sequences (TRSs) upstream to most ORFs in the 3′ one-third of the viral genome (Machhi, 2020). This discontinuous RNA synthesis involves the interaction between complementary TRSs of the nascent negative-strand RNA (negative-sense TRS body) and the positive-strand genomic RNA (positive-sense TRS-L). Upon re-initiation of RNA synthesis at the TRS-L region, a negative-strand copy of the leader sequence is added to the nascent RNA to complete the synthesis of negative-strand sgRNAs (Hu et al., 2021). The negative-strand sgRNAs serve as templates to synthesize a nested set of positive-sense sg mRNAs, which are translated into structural and accessory proteins. Structural proteins, such as the spike (S) protein, envelope (E) protein, membrane (M) protein, and nucleocapsid (N) protein, and interspersed accessory proteins, contribute to virus assembly and budding, with recent evidence suggesting egress via the lysosomal trafficking pathway. Accessory genes, like ORF3a, ORF6, ORF7a, ORF7b, ORF8, and ORF10 in SARS-CoV-2, display high variability and are implicated in modulating host responses to infection and viral pathogenicity, with some functioning as interferon antagonists. These interactions with host proteins and evasion of innate immune responses are critical for successful viral replication and infection. Nonetheless, the precise molecular functions of many accessory proteins remain largely elusive, owing to the absence of homologies to accessory proteins of other coronaviruses or to other known proteins. SARS-CoV-2 assembly occurs within the host's endoplasmic reticulum (ER)-to-Golgi compartment, where viral structural proteins, including spike (S), envelope (E), membrane (M), and nucleocapsid (N) proteins, interact with each other and the viral RNA genome to form new virions. These virions are then transported through the secretory pathway and released from the host cell via exocytosis or, alternatively, through the lysosomal trafficking pathway, which involves interference with lysosomal functions. This intricate process underscores the complex interplay between the virus and host cellular machinery during viral replication and spread.

VITAMINS AND MOLECULAR RELATIONSHIP WITH INFLAMMATION AND INFECTION (IN COVID-19)

The pathogenesis of COVID-19 involves a multifaceted interplay of inflammation, oxidative stress, and immune dysregulation. SARS-CoV-2, primarily transmitted through respiratory droplets and aerosols, manifests a median incubation period of 4–5 days, often preceding symptom onset. Although some cases are asymptomatic, typical presentations include mild to moderate respiratory symptoms such as cough, fever, headache, myalgia, and diarrhea, progressing to severe illness characterized by dyspnea and hypoxemia within approximately 1 week. Severe cases often culminate in acute respiratory distress syndrome (ARDS), marked by severe hypoxemia and pulmonary vascular leakage leading to loss of aerated lung tissue. A hallmark of severe COVID-19 is systemic hyperinflammation, driven by the release of pro-inflammatory cytokines such as interleukin-1 (IL-1), IL-6, IL-8, and Tumour Necrosis Factor (TNF), accompanied by elevated inflammatory markers like D-dimer, ferritin, and C-reactive protein (CRP). This inflammatory cascade contributes to the progression of lung injury, characterized histologically by diffuse alveolar damage (DAD), featuring alveolar epithelial disruption, hyaline membrane formation, and microvascular thrombosis. Notably, COVID-19 patients frequently exhibit a dysregulated coagulation-fibrinolysis balance, resulting in fibrin-rich hyaline membranes and thrombi formation, exacerbating hypoxemia. Immune dysregulation further complicates the pathogenesis, evidenced by immune cell infiltrates, including macrophages and lymphocytes, within lung tissues. Monocytes and macrophages show aberrant activation and dysfunction, accompanied by altered expression of inflammatory mediators. Notably, impaired T-cell responses may contribute to disease severity. Additionally, macrophages may facilitate viral spread within the lungs, while neutrophil enrichment exacerbates respiratory inflammation. Importantly, the gastrointestinal tract can also be infected, contributing to gastrointestinal symptoms and shedding of viral RNA in feces. Considering this complex pathogenesis, interventions targeting inflammation, oxidative stress, and immune dysregulation hold promise. Vitamins, with their antioxidant and immunomodulatory properties, may offer potential benefits. For instance, vitamin C and vitamin D have been studied for their anti-inflammatory effects and potential to mitigate cytokine storms. Vitamin E, with its antioxidant properties, may counteract oxidative stress implicated in lung injury. Furthermore, vitamin A plays a crucial role in maintaining mucosal immunity, potentially aiding in gastrointestinal infection control (Figure 8.2). Since this chapter aims at lipid-soluble vitamins, we will now discuss more on the role of vitamins A, D, E, and K in COVID-19.

Vitamin A, known for its role in maintaining mucosal immunity and epithelial barrier integrity, is crucial in defending against respiratory infections, including those caused by viruses. Adequate vitamin A levels are essential for the production and function of immune cells, particularly T cells, which play a central role in antiviral immune responses. Furthermore, vitamin A contributes to the regulation of inflammatory processes, helping to balance immune responses and mitigate excessive inflammation, which is a hallmark of severe COVID-19. By supporting mucosal immunity and regulating inflammation, vitamin A may contribute to reducing the severity of respiratory infections, including COVID-19.

Vitamin D or calciferol is another highly useful lipid-soluble vitamin, the insufficiency of which has long been recognized for its potential impact on immune function, particularly in the context of viral infections. Vitamin D plays crucial immunomodulatory roles in bolstering the body's defense mechanisms against microbial invaders and pathogens (Ghaseminejad-Raeini et al., 2023). It promotes the synthesis of antimicrobial peptides, pattern recognition receptors (PRRs), and cytokines within cells, thereby enhancing the innate immune response (White, 2022). The vitamin D receptor (VDR), which plays a pivotal role in mediating the hormonal functions across various physiological processes, exhibits widespread expression throughout the body, being present in nearly all organs (Katos, 2000). Its activity occurs through both genomic mechanisms, involving nuclear VDR, and non-genomic pathways, involving membrane-bound VDR. Additionally, vitamin D inhibits the maturation and activation of dendritic cells and hinders the differentiation of monocytes into macrophages, thereby regulating the activity of antigen-presenting cells (Aranow, 2011). In terms of adaptive immunity, vitamin D exerts its influence by modulating T-cell activation and altering the function and phenotype of antigen-presenting cells (Athanassiou et al., 2022). It regulates CD4+ T-cell responses by favoring the development of T helper 2 (Th$_2$) cells while suppressing T helper 1 (Th$_1$) cells, thus influencing the balance of immune responses. Furthermore, vitamin D impacts the phagocytic activity of macrophages and natural killer cells, enhancing their ability to engulf and eliminate pathogens (Hewison, 2010). It also regulates cytokine production, ensuring a balanced immune response without excessive inflammation. Moreover, vitamin D plays a pivotal role in regulating the functions of lymphocytes, contributing to the overall efficiency of the immune system in identifying and neutralizing foreign invaders. Overall, these immunomodulatory properties highlight the importance of vitamin D in maintaining a robust and balanced immune response essential for defending against infections and maintaining overall health. Emerging

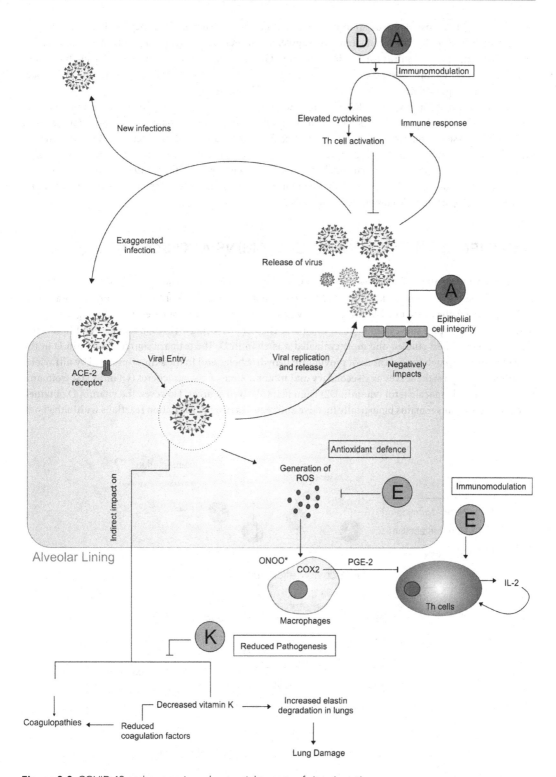

Figure 8.2 COVID-19 pathogenesis and potential targets of vitamin action.

evidence suggests a possible association between vitamin D deficiency and an increased susceptibility to COVID-19. However, the efficacy of vitamin D supplementation in reducing the risk of COVID-19 infection remains uncertain and requires further investigation. Given its safety profile, widespread availability, and cost-effectiveness, vitamin D supplementation holds promise as a potential strategy to attenuate the impact of the COVID-19 pandemic.

Vitamin K, particularly vitamin K2, is recognized for its role in regulating blood coagulation and preventing excessive clotting. COVID-19 patients often exhibit a dysregulated coagulation-fibrinolysis balance, leading to increased thrombotic events and microvascular complications (Meizoso et al., 2021). Vitamin K2, through its involvement in activating proteins that inhibit vascular calcification and promote anticoagulation, may help mitigate the hypercoagulability associated with severe COVID-19. By supporting proper coagulation function and preventing excessive clot formation, vitamin K2 supplementation could potentially reduce the risk of thrombotic complications in COVID-19 patients.

DEFICIENCY OF LIPID-SOLUBLE VITAMINS AND COVID-19

During the COVID-19 pandemic, two types of vitamin deficiencies were observed, one is pre-COVID deficiency and the other is post-lockdown deficiency (Figure 8.3). Irrespective of the nature or temporal cause of the deficiency, the impact of vitamin deficiency has been negative in most cases in increasing morbidity. Based on our literature survey, until when this book chapter was being written, among several lipid-soluble vitamins, the most studied and most evaluated was vitamin D. The primary source of vitamin D in the human body is through the synthesis of provitamin dehydrocholesterol in the skin, stimulated by ultraviolet radiation. Dietary intake serves as a secondary and minor source of cholecalciferol (vitamin D3) from animal products and ergocalciferol (vitamin D2) from plant-derived sources. However, the vitamin D obtained through these means remains biologically inactive and requires two hydroxylation reactions within the body

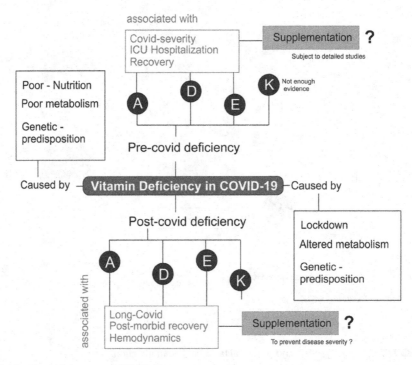

Figure 8.3 Vitamin deficiency and COVID-19.

for activation. The initial hydroxylation takes place in the liver, converting vitamin D into 25-hydroxyvitamin D [25(OH)D], also referred to as calcidiol. Subsequently, the second hydroxylation primarily occurs in the kidney, yielding the biologically active form, 1,25-dihydroxyvitamin D [1,25(OH)2D], commonly known as calcitriol (Ross et al., 2011). Evaluating this mechanism in a retrospective manner, when lockdown was imparted, researchers observed a general lack of sun exposure and hence likely reduction in vitamin D synthesis. A clinical study by Chandankere et al. (2023) showed an observable statistically significant rise in symptomatic vitamin D deficiency (VDD) among adolescents during the 6 months following the lockdown period. The study involved the examination of 31 adolescents aged between 9 and 14 years. Within a span of 12 weeks, all individuals experienced relief from pain, and their muscle strength returned to normal levels. This was further alleviated with supplemental vitamin D (Chandankere et al., 2023). Furthermore, meta-analysis studies revealed and emphasized the fact that VDD may increase the risk of COVID-19 infection and the likelihood of severe disease (Kaya et al., 2021; Pereira et al., 2022; Dantas Damascena et al., 2023). Besides vitamin D, some studies also showed an association between vitamins A and E with COVID. A case-control study by Manduor et al. deciphered that the levels of vitamin A were reduced in COVID-19 patients particularly in the intensive care unit (ICU). This ensures the association of decreased vitamin A with disease morbidity and the importance of vitamin A supplementation as part of disease management (Mandour et al., 2023). In yet another interesting concept, the prevalence of long COVID has been associated with COVID-19. Long COVID is a term defined by CDC as "Prevalence and persistence of signs, symptoms, and conditions that continue or develop after initial SARS-CoV-2 infection for a period beyond four weeks or more after the initial phase of infection; may be multisystemic; and may present with a relapsing–remitting pattern and progression or worsen over time, with the possibility of severe and life-threatening events even months or years after infection" (CDC). Fillipo et al. showed that COVID-19 survivors with long COVID have lower 25(OH) vitamin D levels than matched patients without long COVID. The authors of the study also suggested that vitamin D levels should be evaluated in COVID-19 patients after hospital discharge (di Filippo et al., 2023). In an Italian retrospective study, a significantly elevated prevalence of hypovitaminosis D was observed in hospitalized patients with COVID-19, indicating a potential link between low vitamin D levels and an increased susceptibility to SARS-CoV-2 infection and subsequent hospitalization. This cohort study also revealed a reverse correlation between serum 25(OH)D levels and the risk of in-hospital mortality, implying that a lower vitamin D status upon admission could serve as a modifiable and autonomous risk factor for unfavorable outcomes in COVID-19 cases (Infante et al., 2022). Detection of early precursors of vitamin D and associating the link with COVID-19 was yet another approach followed in some clinical studies, e.g., a retrospective study by Infante in 2021 demonstrated that reduced total testosterone levels and increased E2/T (estradiol/testosterone) ratio upon admission correlate with a hyperinflammatory condition in male COVID-19 patients during hospitalization. Also, they showed that diminished total testosterone levels upon admission independently predict in-hospital mortality in the group, suggesting the utility of testosterone levels and the E2/T ratio as prognostic indicators of disease severity among these individuals (Infante et al., 2021). In a cohort study (n = 150) by Sulli et al., the COVID-19 cohort exhibited notably lower levels of vitamin D compared to the healthy control. Severe VDD was identified in 24.0% of COVID-19 patients versus 7.3% in the control group, suggesting that deficiencies in vitamin D might compromise immune defenses against COVID-19, potentially leading to disease exacerbation (Sulli et al., 2021). In an observation study of 149 patients in Turkey, Serum 25(OH) vitamin D was independently associated with mortality in COVID-19 patients (Karahan & Katkat, 2021). In yet another observational study, the ICU patients exhibited a diminished serum concentration of 25-hydroxyvitamin D (25(OH)D). The authors of the study projected that a serum concentration of 25(OH)D ≤ 9.9 ng/mL upon admission can serve as a predictive marker for in-hospital mortality among COVID-19 patients (Bychinin et al., 2021). In a retrospective study aimed at examination of the markers of vitamin D status and degradation products of vitamin D in a diverse cohort of 148 COVID-19 patients hospitalized with different clinical presentations of the disease, survivors and non-survivors did not show a significant difference in serum concentrations of 25(OH)D vitamin D and the two vitamin D catabolites (Zelzer et al., 2021).

VITAMIN SUPPLEMENTATION IN COVID-19 – CLINICAL STUDIES

Recommended doses of vitamins A, D, E, and K vary depending on factors such as age, sex, and individual health conditions. However, generally accepted guidelines suggest daily intake levels of 900 micrograms (mcg) for men and 700 mcg for women of vitamin A, primarily obtained from dietary sources such as liver, fish, and dairy products. For vitamin D, the recommended daily allowance is 600 international units (IU) for individuals aged 1–70 years and 800 IU for those over 70 years, although some experts propose higher doses for specific populations. Vitamin E intake recommendations range from 15 milligrams (mg) to 1,000 mg per day, with a typical supplemental dose around 15–100 mg. Vitamin K guidelines recommend 120 mcg per day for men and 90 mcg for women, primarily from green leafy vegetables and certain oils (Schwartz, 2001). While adequate levels of these vitamins are essential for overall health, their potential benefits in preventing and reducing the severity of COVID-19 remain under investigation. Preliminary studies suggest that optimizing levels of these vitamins through supplementation may support immune function and reduce inflammation, thus potentially lowering the risk of contracting COVID-19 or experiencing severe symptoms. There have been a number of randomized controlled clinical trials and meta-analysis demonstrating the benefits of vitamin supplementation in either preventing the exaggeration of symptoms of COVID-19 or increased recovery. Also, some studies have also been performed on the analysis of risk reduction toward COVID-19 infection by lipid-soluble vitamin supplementation. Some of the notable studies are listed in Table 8.1. We will discuss here both positive and neutral study outcomes on vitamin supplementation. Readers may however consider the fact that by the time the text was written, most of the studies were performed on vitamin D, while seldom evidences were found on supplementation of other lipid-soluble vitamins such as vitamins A, E, and K.

Table 8.1 Some notable clinical studies on lipid-soluble vitamin supplementation in COVID-19 and their outcomes

	Study title	Authors	Nature of study	Sample size	Outcome
1	Vitamin D supplementation and COVID-19 (ICU)	Domazet Bugarin et al. (2023)	Randomized controlled trial	155	No benefit in vitamin D supplementation to patients with severe COVID-19 disease admitted to the ICU
2	Calcidiol in the cholecalciferol	Cannata-Andía et al. (2022)	Randomized controlled trial	274	Did not improve the outcomes of the COVID-19 disease
3	Vitamin D supplementation in frontline health workers	Villasis-Keever et al. (2022)	Randomized controlled trial	321	Supplementation in highly exposed individuals prevents SARS-CoV-2 infection without serious AEs
4	Effect of a test-and-treat approach to vitamin D supplementation	Jolliffe et al. (2022) (Trial Registration no. NCT04579640)	Phase 3 randomized controlled trial	1,600	Population-level test-and-treat approach to vitamin D supplementation was not associated with a reduction in risk
5	Effect of vitamin D supplementation on COVID-19-related intensive care hospitalization and mortality	Argano et al. (2023)	Meta-analysis & trial sequential analysis	78	Vitamin D administration results in a decreased risk of death and ICU admission

(Continued)

Table 8.1 (*Continued*) Some notable clinical studies on lipid-soluble vitamin supplementation in COVID-19 and their outcomes

	Study title	Authors	Nature of study	Sample size	Outcome
6	Comparative effects of two vitamin D3 doses on recovery of symptoms	Sabico et al. (2021)	Randomized Controlled Trial	69	A 5000 IU daily oral vitamin D3 supplementation for 2 weeks reduces the time to recovery for cough and gustatory sensory loss among patients with sub-optimal vitamin D status and mild to moderate COVID-19 symptoms.
7	Efficacy of a high dose of vitamin D on the hyperinflammation state in moderate to severe COVID-19 patients	Sarhan et al. (2022)	Randomized controlled trial	116	High-dose vitamin D was considered a promising treatment in the suppression of cytokine storms among COVID-19 patients and was associated with better clinical improvement
8	Effect of vitamin D supplementation on disease progression and post-exposure prophylaxis for COVID-19	Wang et al. (2021)	Cluster-randomized design	2,700	Cost-efficient approach to testing an intervention for reducing rates of hospitalization and/or mortality in newly diagnosed cases and preventing infection
9	Efficacy of vitamin A as an adjuvant therapy for pneumonia in children	Li et al. (2022)	Meta-analysis	3,496	Vitamin A contributes to relieve the clinical symptoms and signs and also shortens the hospitalization period.
10	Comparison of vitamin A supplementation with standard therapies	Rohani et al. (2022)	Randomize clinical trial	182	Vitamin A supplementation demonstrated efficacy in improving some clinical and paraclinical symptoms in patients with COVID-19

Recently, Gibbons et al. investigated whether vitamin D supplementation could reduce the risk of COVID-19 infection in individuals with low levels of vitamin D. While previous research has linked VDD with weakened immune function and increased susceptibility to viral infections, including COVID-19. This study in a cohort of US veterans demonstrated that supplementation with vitamins D2 and D3 was associated with a reduction in COVID-19 infection rates. Specifically, vitamin D3 supplementation led to a 28% decrease, while vitamin D2 led to a 20% decrease in infection rates. Moreover, mortality within 30 days of COVID-19 infection was lower among individuals supplemented with vitamin D3 (33% decrease) and D2 (25% decrease). Interestingly, the study also observed that veterans with lower baseline vitamin D levels experienced greater benefits from supplementation (Gibbons et al., 2022). While investigators further looked at the genetic level and possible explanation of the vitamin D-associated risk prevention, they observed interesting results. In a study aimed to assess the prevalence of high-impact variants in genes associated with the vitamin D pathways in the Portuguese population and their relationship with COVID-19 outcomes resulted in an observed association between the vitamin D polygenic risk score and serum 25(OH)D concentration. In the study by Freitas et al, a total of 517 patients admitted to two tertiary Portuguese hospitals were enrolled, and their serum concentration of 25-hydroxyvitamin D (25(OH)D) was measured upon admission. Genetic analysis was conducted for 18 variants in genes, including AMDHD1, CYP2R1, CYP24A1, DHCR7, GC, SEC23A, and VDR. The findings revealed that polymorphisms in the vitamin D-binding protein encoded by the GC gene were significantly associated with infection severity. Importantly, serum 25(OH)D levels were also associated with survival outcomes, indicating a potential role of vitamin D in COVID-19 prognosis. Furthermore, the Portuguese population exhibited a higher prevalence of the DHCR7 RS12785878 variant compared to the European population, suggesting a genetic susceptibility to VDD that may contribute to the severity of COVID-19 (Freitas et al., 2021). It is well-established that vitamin D serves as an immunomodulatory molecule, which has also been described in other chapters of this book; the exact mechanism of immunomodulation in COVID-19 and thereby its prophylactic role is an open question to the infection biology researchers. In an attempt, Hafezi et al. investigated the effects of vitamin D (VitD) on host interferon-alpha/beta (IFN-α/β) signaling in severe COVID-19 patients using blood and saliva samples from 43 severe COVID-19 patients treated with VitD in contrast with 37 untreated patients in the ICU with a follow-up to 29 days post-admission. The study revealed a heightened activity level of RIG-1/MDA-5 and JAK-STAT signaling pathways, along with increased expression of antiviral interferon-stimulating genes (ISGs) such as MX-1 and ISG-15 in vitro in peripheral blood mononuclear cells (PBMCs) treated with VitD, as well as in blood and saliva samples from VitD-treated patients. Furthermore, VitD-treated patients exhibited a reduced risk of mortality compared to untreated patients by day 29 post-admission (Hafezi et al., 2022). In another study evaluating the immunomodulatory outcomes in COVID-19 patients supplemented with vitamin D, admitted to ICU, interesting results were obtained. Patients receiving vitamin D3 exhibited notably higher NK and NKT cell counts and NLR compared to those receiving placebo on day 7. However, there were no significant differences observed in mortality rates (37% vs 50%, $P = 0.16$), need for mechanical ventilation (63% vs 69%, $P = 0.58$), or the incidence of nosocomial infection (60% vs 41%, $P = 0.05$) between the two groups. Although vitamin D3 supplementation led to increased lymphocyte counts. At the same time, the study also concluded no reduction in mortality among patients in the ICU when compared to placebo (Bychinin et al., 2022).

VITAMIN SUPPLEMENTATION IN COVID-19 AND NEGATIVE OR NEUTRAL IMPACTS – CLINICAL STUDIES

While we witnessed most of the studies indicating the benefits of lipid-soluble vitamins as supplements in alleviating the symptoms or severity, some studies have also indicated the contrary. Researchers also noticed some neutral outcomes, i.e., no effect of vitamin supplementation during controlled randomized trials (Table 8.1). In a pilot randomized controlled trial, Somi et al. have shown that there is no advantage of vitamin A supplementation when compared to the standard treatment in determining the severity of outcomes in COVID-19 patients admitted to hospitals (Somi et al., 2022). A study by Bugarin et al. aimed to assess the impact of daily vitamin D supplementation during ICU stay on 155 COVID-19 patients with severe illness. The trial found no significant difference in the duration of respiratory support between the two groups, although the study lacked sufficient power for this outcome. Additionally, there were no disparities observed in any of the secondary outcomes analyzed (Domazet Bugarin et al., 2023). Additionally, a clinical trial found that giving a large dose of 100,000 IU of cholecalciferol by mouth when patients were admitted to the hospital didn't make COVID-19 outcomes better (Cannata-Andía et al., 2022). Some studies have also shown the ill effects of high doses of fat-soluble vitamins. The safety of vitamin D supplementation remains consistent across doses of 400, 4,000, and 10,000 IU per day. Hypercalciuria is a common occurrence, with higher doses showing a higher frequency and more likely with increased doses (Billington et al., 2020). While important for overall health, excessive intake of preformed vitamin A can lead to acute and chronic toxicity, known as hypervitaminosis A. Vitamin A is also considered a teratogen, capable of causing severe birth defects (Olson et al., 2022). Nevertheless, researchers have also looked forward to carry out mega-dose vitamin A clinical trials in COVID-19 (Midha et al., 2021).

Some clinical trials are still ongoing and the outcome is still awaited. Clinicaltrials.gov does not currently list any trials investigating the sole use of vitamin E in COVID-19 patients, but there are ongoing studies utilizing vitamin E in conjunction with other compounds. For instance, a clinical trial conducted in Spain is exploring the efficacy of a micronutrient supplement containing 45 mg of vitamin E (as alpha-tocopherol) along with ten additional vitamins and minerals over a 14-day period, aiming to reduce the need for hospitalization in 300 outpatient adults diagnosed with COVID-19 (CoviT-1). Similarly, another trial in Saudi Arabia is assessing the impact of a dietary supplement consisting of 90 mg of vitamin E (specific form not specified), alongside 1,500 mcg of vitamin A (as beta-carotene), 250 mg of vitamin C, 15 mcg of selenium, and 7.5 mg of zinc over a 14-day period on the disease progression and the risk of cytokine storm in 40 adults diagnosed with COVID-19 (ONSCOVID). Despite these findings being unfavorable, there remains a pressing requirement for forthcoming clinical investigations to offer a more substantial level of confirmation.

CONCLUSION

Despite numerous studies and ongoing clinical and basic research, we still face a critical clinical question: should lipid-soluble vitamins be prescribed for infectious diseases like COVID-19? If so, what criteria should guide supplementation? Should serum levels of fat-soluble vitamins be used as prognostic markers and a basis for supplementation? Furthermore, if the answer is affirmative, what would be the optimal dosage? Conversely, the answer could also be negative, with scientific justification still lacking. Additionally, there is insufficient evidence to endorse or discourage the use of any dietary supplement, including vitamins, minerals, herbs, botanicals, fatty acids, or other dietary ingredients, for the prevention or treatment of COVID-19. It is essential to note that, according to regulations, dietary supplements cannot be promoted as treatments, preventions, or cures for any disease; only pharmaceutical drugs can make such claims. Nevertheless, sales of dietary supplements marketed for immune support surged following the onset of COVID-19, as many individuals hoped these products might offer some defense against SARS-CoV-2 infection and potentially mitigate the severity of the disease for those affected. We, as authors and biochemists, also advocate for the emerging concept of precision vitamin supplementation in light of advancements in genomic fingerprinting and the availability of high-throughput metabolomics and genomics approaches to answer the aforementioned questions.

REFERENCES

Aranow, C. (2011, August). Vitamin D and the immune system. *Journal of Investigative Medicine, 59*(6), 881–886.

Argano, C., Mallaci Bocchio, R., Natoli, G., Scibetta, S., Lo Monaco, M., & Corrao, S. (2023). Protective effect of vitamin D supplementation on COVID-19-related intensive care hospitalization and mortality: definitive evidence from meta-analysis and trial sequential analysis. *Pharmaceuticals, 16*(1), 130.

Athanassiou, L., Mavragani, C. P., & Koutsilieris, M. (2022, March 31). The immunomodulatory properties of vitamin D. *Mediterranean Journal of Rheumatology, 33*(1), 7–13.

Billington, E. O., Burt, L. A., Rose, M. S., Davison, E. M., Gaudet, S., Kan, M.,... & Hanley, D. A. (2020). Safety of high-dose vitamin D supplementation: secondary analysis of a randomized controlled trial. *The Journal of Clinical Endocrinology & Metabolism, 105*(4), 1261–1273.

Bychinin, M. V., Klypa, T. V., Mandel, I. A., Andreichenko, S. A., Baklaushev, V. P., Yusubalieva, G. M.,... & Troitsky, A. V. (2021). Low circulating vitamin D in intensive care unit-admitted COVID-19 patients as a predictor of negative outcomes. *The Journal of Nutrition, 151*(8), 2199–2205.

Bychinin, M. V., Klypa, T. V., Mandel, I. A., Yusubalieva, G. M., Baklaushev, V. P., Kolyshkina, N. A., & Troitsky, A. V. (2022). Effect of vitamin D3 supplementation on cellular immunity and inflammatory markers in COVID-19 patients admitted to the ICU. *Scientific Reports*, *12*(1), 18604.

Cannata-Andía, J. B., Díaz-Sottolano, A., Fernández, P., Palomo-Antequera, C., Herrero-Puente, P., Mouzo, R., ... & COVID-VIT-D Trial Collaborators. (2022). A single-oral bolus of 100,000 IU of cholecalciferol at hospital admission did not improve outcomes in the COVID-19 disease: the COVID-VIT-D-a randomised multicentre international clinical trial. *BMC Medicine*, *20*(1), 83.

CDC. Long COVID or Post-COVID Conditions. https://www.cdc.gov/coronavirus/2019-ncov/long-term-effects/index.html

Chandankere, V., Konanki, R., Reddy Maryada, V., & Reddy, A. G. (2023). Impact of COVID lockdown: increased prevalence of symptomatic vitamin D deficiency in adolescents. *Journal of Clinical Orthopaedics and Trauma*, *47*, 102316.

Dantas Damascena, A., Galvão Azevedo, L. M., de Almeida Oliveira, T., da Mota Santana, J., & Pereira, M. (2023). Vitamin D deficiency aggravates COVID-19: discussion of the evidence. *Critical Reviews in Food Science and Nutrition*, *63*(4), 563–567.

di Filippo, L., Frara, S., Nannipieri, F., Cotellessa, A., Locatelli, M., Rovere Querini, P., & Giustina, A. (2023). Low vitamin D levels are associated with Long COVID syndrome in COVID-19 survivors. *The Journal of Clinical Endocrinology and Metabolism*, *108*(10), e1106–e1116.

Domazet Bugarin, J., Dosenovic, S., Ilic, D., Delic, N., Saric, I., Ugrina, I., ... & Saric, L. (2023). Vitamin D supplementation and clinical outcomes in severe COVID-19 patients-randomized controlled trial. *Nutrients*, *15*(5), 1234.

Freitas, A. T., Calhau, C., Antunes, G., Araújo, B., Bandeira, M., Barreira, S., ... & Pinto, F. J. (2021). Vitamin D-related polymorphisms and vitamin D levels as risk biomarkers of COVID-19 disease severity. *Scientific Reports*, *11*(1), 20837.

Ghaseminejad-Raeini, A., Ghaderi, A., Sharafi, A., Nematollahi-Sani, B., Moossavi, M., Derakhshani, A., & Sarab, G. A. (2023, July 14). Immunomodulatory actions of vitamin D in various immune-related disorders: a comprehensive review. *Frontiers in Immunology*, *14*, 950465.

Gibbons, J. B., Norton, E. C., McCullough, J. S., Meltzer, D. O., Lavigne, J., Fiedler, V. C., & Gibbons, R. D. (2022). Association between vitamin D supplementation and COVID-19 infection and mortality. *Scientific Reports*, *12*(1), 19397.

Hafezi, S., Saheb Sharif-Askari, F., Saheb Sharif-Askari, N., Ali Hussain Alsayed, H., Alsafar, H., Al Anouti, F., ... & Halwani, R. (2022). Vitamin D enhances type I IFN signaling in COVID-19 patients. *Scientific Reports*, *12*(1), 17778.

Hewison, M. (2010, June), Vitamin D and the immune system: new perspectives on an old theme. *Endocrinology and Metabolism Clinics of North America*, *39*(2), 365–379.

Hu, B., Guo, H., Zhou, P. & Shi, Z.-L. (2021). Characteristics of SARS-CoV-2 and COVID-19. *Nature Reviews. Microbiology*, *19*, 141–154.

Infante, M., Buoso, A., Pieri, M., Lupisella, S., Nuccetelli, M., Bernardini, S., & Morello, M. (2022, March–April). Low vitamin D status at admission as a risk factor for poor survival in hospitalized patients with COVID-19: an Italian retrospective study. *Journal of the American Nutrition Association*, *41*(3), 250–265.

Infante, M., Pieri, M., Lupisella, S., D'Amore, L., Bernardini, S., Fabbri, A., & Morello, M. (2021, October). Low testosterone levels and high estradiol to testosterone ratio are associated with hyperinflammatory state and mortality in hospitalized men with COVID-19. *European Review for Medical and Pharmacological Sciences*, *25*(19), 5889–5903. doi: 10.26355/eurrev_202110_26865.

Jackson, C. B., Farzan, M., Chen, B., & Choe, H. (2022). Mechanisms of SARS-CoV-2 entry into cells. *Nature Reviews Molecular Cell Biology*, *23*(1), 3–20.

Jolliffe, D. A., Holt, H., Greenig, M., Talaei, M., Perdek, N., Pfeffer, P., ... & Martineau, A. R. (2022). Effect of a test-and-treat approach to vitamin D supplementation on risk of all cause acute respiratory tract infection and covid-19: phase 3 randomised controlled trial (CORONAVIT). *BMJ*, *378*, e071230.

Karahan, S., & Katkat, F. (2021). Impact of serum 25 (OH) vitamin D level on mortality in patients with COVID-19 in Turkey. *The Journal of Nutrition, Health & Aging*, *25*(2), 189–196.

Kato, S. (2000, May). The function of vitamin D receptor in vitamin D action. *Journal of Biochemistry, 127*(5), 717–722.

Kaya, M. O., Pamukçu, E., & Yakar, B. (2021). The role of vitamin D deficiency on COVID-19: a systematic review and meta-analysis of observational studies. *Epidemiology and Health, 43*, e2021074.

Li, R., Zhao, W., Wang, H., Toshiyoshi, M., Zhao, Y., & Bu, H. (2022). Vitamin A in children's pneumonia for a COVID-19 perspective: a systematic review and meta-analysis of 15 trials. *Medicine, 101*(42), e31289.

Machhi, J., Herskovitz, J., Senan, A. M., Dutta, D., Nath, B., Oleynikov, M. D., ... & Kevadiya, B. D. (2020, September). The natural history, pathobiology, and clinical manifestations of SARS-CoV-2 infections. *Journal of Neuroimmune Pharmacology, 15*(3), 359–386.

Mandour, I. A., Hussein, S. A., Hanna, H. W. Z., Abdellatif, S. A., & Fahmy, B. S. (2023). Evaluation of vitamin A and E deficiency with severity of SARS-COV-2 disease: a case-control study. *The Egyptian Journal of Bronchology, 17*(1), 36.

Meizoso, J. P., Moore, H. B., & Moore, E. E. (2021, June). Fibrinolysis shutdown in COVID-19: clinical manifestations, molecular mechanisms, and therapeutic implications. *Journal of the American College of Surgeons, 232*(6), 995–1003.

Midha, I. K., Kumar, N., Kumar, A., & Madan, T. (2021). Mega doses of retinol: a possible immunomodulation in Covid-19 illness in resource-limited settings. *Reviews in Medical Virology, 31*(5), 1–14.

Olson, J. M., Ameer, M. A., & Goyal, A. (2022). *Vitamin A Toxicity*. In StatPearls [Internet]. StatPearls Publishing.

Pereira, M., Dantas Damascena, A., Galvão Azevedo, L. M., de Almeida Oliveira, T., & da Mota Santana, J. (2022). Vitamin D deficiency aggravates COVID-19: systematic review and meta-analysis. *Critical Reviews in Food Science and Nutrition, 62*(5), 1308–1316.

Rohani, M., Mozaffar, H., Mesri, M., Shokri, M., Delaney, D., & Karimy, M. (2022). Evaluation and comparison of vitamin A supplementation with standard therapies in the treatment of patients with COVID-19. *Eastern Mediterranean Health Journal, 28*(9), 673–681.

Ross, A. C., Taylor, C. L., Yaktine, A. L., & Del Valle, H. B. (2011). Committee to review dietary reference intakes for vitamin D and calcium. Food and Nutrition Board. National Academies Press. https://www.ncbi.nlm.nih.gov/books/NBK56070/

Sabico, S., Enani, M. A., Sheshah, E., Aljohani, N. J., Aldisi, D. A., Alotaibi, N. H.,... & Al-Daghri, N. M. (2021). Effects of a 2-week 5000 IU versus 1000 IU vitamin D3 supplementation on recovery of symptoms in patients with mild to moderate covid-19: a randomized clinical trial. *Nutrients, 13*, 2170.

Sarhan, N., Abou Warda, A. E., Sarhan, R. M., Boshra, M. S., Mostafa-Hedeab, G., ALruwaili, B. F.,... & Fathy, S. (2022). Evidence for the efficacy of a high dose of vitamin D on the hyperinflammation state in moderate-to-severe COVID-19 patients: a randomized clinical trial. *Medicina, 58*(10), 1358.

Schwartz, J. B. (2001). Vitamin intake, recommended intake, and gender differences. *The Journal of Gender-Specific Medicine: JGSM: The Official Journal of the Partnership for Women's Health at Columbia, 4*(1), 11–15.

Shirbhate, E., Pandey, J., Patel, V. K., Kamal, M., Jawaid, T., Gorain, B.,... & Rajak, H. (2021). Understanding the role of ACE-2 receptor in pathogenesis of COVID-19 disease: a potential approach for therapeutic intervention. *Pharmacological Reports*, 1–12.

Somi, M. H., Faghih Dinevari, M., Taghizadieh, A., Varshochi, M., Sadeghi Majd, E., Abbasian, S., & Nikniaz, Z. (2022). Effect of vitamin A supplementation on the outcome severity of COVID-19 in hospitalized patients: a pilot randomized clinical trial. *Nutrition and Health*, 02601060221129144. doi:10.1177/02601060221129144

Sulli, A., Gotelli, E., Casabella, A., Paolino, S., Pizzorni, C., Alessandri, E.,... & Cutolo, M. (2021). Vitamin D and lung outcomes in elderly COVID-19 patients. *Nutrients, 13*(3), 717.

Valyaeva, A. A., Zharikova, A. A., & Sheval, E. V. (2023). SARS-CoV-2 cellular tropism and direct multiorgan failure in COVID-19 patients: bioinformatic predictions, experimental observations, and open questions. *Cell Biology International, 47*(2), 308–326.

Villasis-Keever, M. A., López-Alarcón, M. G., Miranda-Novales, G., Zurita-Cruz, J. N., Barrada-Vázquez, A. S., González-Ibarra, J.,... & Parra-Ortega, I. (2022). Efficacy and safety of vitamin D supplementation to prevent COVID-19 in frontline healthcare workers. A randomized clinical trial. *Archives of Medical Research, 53*(4), 423–430.

V'kovski, P., Kratzel, A., Steiner, S., Stalder, H., & Thiel, V. (2021). Coronavirus biology and replication: implications for SARS-CoV-2. *Nature Reviews Microbiology*, *19*(3), 155–170.

White, J. H. (2022, January 11). Emerging roles of vitamin D-induced antimicrobial peptides in antiviral innate immunity. *Nutrients*, *14*(2), 284.

WHO. (2020). Coronavirus Disease (COVID-19) Pandemic. https://www.who.int/emergencies/diseases/novel-coronavirus-2019- (Accessed on 23 December 2023).

Yadav, R., Chaudhary, J. K., Jain, N., Chaudhary, P. K., Khanra, S., Dhamija, P.,... & Handu, S. (2021). Role of structural and non-structural proteins and therapeutic targets of SARS-CoV-2 for COVID-19. *Cells*, *10*(4), 821.

Yan, W., Zheng, Y., Zeng, X., He, B., & Cheng, W. (2022). Structural biology of SARS-CoV-2: open the door for novel therapies. *Signal Transduction and Targeted Therapy*, *7*(1), 26.

Wang, R., DeGruttola, V., Lei, Q., Mayer, K. H., Redline, S., Hazra, A.,... & Manson, J. E. (2021). The vitamin D for COVID-19 (VIVID) trial: a pragmatic cluster-randomized design. *Contemporary Clinical Trials*, *100*, 106176.

Zelzer, S., Prüller, F., Curcic, P., Sloup, Z., Holter, M., Herrmann, M., & Mangge, H. (2021). Vitamin D metabolites and clinical outcome in hospitalized COVID-19 patients. *Nutrients*, *13*(7), 2129.

9

Preventive and therapeutic roles of water-soluble vitamins in Ebola infection

AYUSHI SHARMA AND ANJANA GOEL

INTRODUCTION

There are approximately 1500 species of infectious organisms that can harm people, including 217 viruses and prions [1,2]. The two main agents of unpredictable hemorrhagic fever disease outbreaks are the Marburg virus, which has a 90% mortality rate, and various Ebola virus species [3]. The ongoing global Ebola outbreak was of an unprecedented scope and posed a serious threat to the countries of West Africa and beyond. Ebola and other enveloped viruses are members of the family Filoviridae. It contains one Marburg virus strain and five distinct Ebola species that inflict hemorrhagic fever on both primates and non-primate hosts. The term "viral hemorrhagic fever" refers to a variety of viral illnesses that make people bleed and have a fever. This syndrome is typically brought on by RNA viruses from the families Flaviviridae, Filoviridae, Bunyaviridae, and Arenaviridae [4,5]. There were major hemorrhagic fever cases with a high mortality rate in the Sudanese and Zairean provinces. The causing agents were closely related to the Marburg virus strain, also known as the Ebola virus, and isolated from Germany in 1967 and South Africa in 1975, despite sharing many morphological traits with it [6]. This was confirmed by electron microscopy and serological studies. The Zaire Ebola virus (ZEBOV) was identified in 1976 and holds the record for the highest case fatality rates, with approximately 88% of cases resulting in death, and death rates of around 53%. This virus has predominantly affected regions in Black and West Africa, leading to significant epidemics characterized by high mortality rates. Despite the recurring occurrence of Ebola outbreaks, there is currently no effective vaccine or treatment available for clinical use.

ORIGIN AND DISTRIBUTION

Nonhuman primates and pigs are linked to one Reston Ebola virus (REBOV), which was isolated in the Philippines. The four human-infectious Ebola virus species that are known to originate from Africa are the Tai Forest virus (TAFV), the Bundibugyo Ebola virus (BEBOV), the Sudan Ebola virus (SEBOV), and the Zaire Ebola virus (ZEBOV) [7]. There are five species in the genus Ebola virus. Although it's believed that this strain isn't harmful to people, it can cause hemorrhagic fever in experimentally infected animals [8,9]. The only zoonotic transmission that caused the current Ebola virus outbreak in West Africa, according to reports, was made by a two-year-old boy in Meliandou, Guinea [10].

GENERAL MECHANISM AND EPIDEMIOLOGY OF EBOLA

Considerable quantities of Ebola virus particles have been detected within the sweat glands of human skin, indicating the possibility of transmission through direct contact, even though the exact mechanism of viral entry into the body remains unclear. Human-to-human transmission is commonly observed when unsensitized medical equipment is utilized during treatment and during traditional funeral practices such as physical contact, kissing, and bathing of the deceased [11]. In the wild or in laboratory settings, there hasn't been any proof of transmission through the air between primates, including humans. Unlike other viruses, the

DOI: 10.1201/9781003435686-9

Ebola virus exhibits its highest transmissibility during the clinical phase of infection [12]. This unique characteristic, combined with modern transportation systems, enables the virus to potentially spread to any location worldwide. Research suggests that the drier conditions toward the conclusion of the rainy season could facilitate the transmission of the virus from its concealed reservoir to humans, potentially triggering the spread of Ebola virus disease (EVD) [12].

REPLICATION AND PATHOGENICITY OF EBOLA VIRUS

The Zaire Ebola virus (ZEBOV) is capable of replication, affects pigs, and propagates to other animals, just like the Reston Ebola virus (REBOV). Experiments were conducted to determine the way virus spreads from contaminated to naive animals [13]. The findings demonstrated that the virus replicated in the respiratory system, and transmission was proven in every pig that was naive. EBOV proteins interact with a variety of host proteins to aid in virus replication. Numerous host proteins have been identified as crucial cellular players in the EBOV entry step [14–18], including cathepsin L/B, Mer, mucin domain 1, Niemann-Pick C1, Tyro3 receptor tyrosine kinase family Axl, T-cell immunoglobulin, and cathepsin L/B. The first cellular element that regulates the replication and transcription processes of the EBOV genome is DNA topoisomerase 1 (TOP1), but the majority of the cellular host elements are still unknown [19].

MODE OF ACTION

The EBOV-glycoprotein (EBOV-GP), responsible for facilitating viral entry and promoting viral release from host cells, has a molecular weight of approximately 140 kDa [20]. EBOV-GP is initially synthesized as single-chain precursors, which are then co-translated within the endoplasmic reticulum (ER) lumen to form trimers [21]. To initiate infection and progression, the receptor-binding subunit of GP undergoes post-translational cleavage, resulting in the formation of two separate chains, GP1 and GP2 [22]. The cleavage of GP by proteases is linked to the pH-dependent entry of the virus [23]. Cleavage generated three fragments with masses of 23, 19, and 4 kDa [24]. A 19-kDa core subunit potentiates GP's capacity to undergo subsequent conformational changes necessary for fusion when it binds endosomal receptors [25,26]. The binding to receptors is primarily governed by the GP1 subunit, whereas the GP2 subunit serves to anchor the glycoprotein within the viral membrane. The initial 300 residues of the GP1 subunit exhibit conservation, whereas the remaining residues display variability. Toward the C-terminus of GP1, the mucin domain represents a region of variation, characterized by numerous O-linked glycosylation sites [27,28].

A structural analysis of GP2 using X-ray crystallography revealed that the protein consists of a central triple-stranded coiled coil, followed by a disulfide-bonded loop. This configuration suggests that GP2 may serve as a bridge between two membranes to initiate membrane fusion [29]. Furthermore, the trimeric crystal structure of the surface glycoprotein (GP) is essential for the attachment and fusion of viral and host membranes. This process is further facilitated by the binding of a neutralizing antibody, KZ52, in humans. Niemann-Pick C1 (NPC1), a cholesterol transporter found in human lysosomes, fulfills a crucial role as a viral receptor by selectively binding to the viral GP [30,31]. The pathogenicity of the Ebola virus, which leads to cytotoxic effects in human endothelial cells, is primarily influenced by the EBOV-GP, representing the key viral factor [30]. The presence of a mucin-like domain within the GP resulted in a significant loss of endothelial cells and increased vascular permeability in explanted human or porcine blood vessels, with effects lasting for 24 hours. This phenomenon was specifically attributed to the mucin-like domain of GP, signifying its role as a viral determinant of Ebola pathogenicity [30]. For EBOV-GP-dependent entry, another AMPK component in macropinocytic events was also required [31]. To avoid being neutralized by antibodies and proteases, these filoviruses continue to maintain glycoprotein glycosylation at the expense of effective entry. The crucial role of enzymes such as cathepsin B (CatB) and cathepsin L (CatL) is observed in the proteolytic cleavage of the EBOV-GP subunit, GP1, which leads to a significant reduction in the multiplication of infectious ZEBOV [32].

VITAMINS B AND C AND THEIR ROLES IN GENERAL METABOLISM

There is a bidirectional correlation between malnutrition and infection. Various stages of immune responses require different micronutrients, including vitamins A, D, C, E, B6, B12, folate, zinc, iron, copper, and selenium. For instance, vitamin A plays a critical role in maintaining the structure and function of skin and mucosal cells, regulating interleukin-2 (IL-2) and TNF levels, and activating antimicrobial activity in macrophages [33,34]. Vitamins D, A, B6, B12, and folate, along with the balance between commensal and pathogenic microorganisms, influence the composition of the intestinal microbiota [35,36]. In this way, a balance between innate and adaptive immune homeostasis allows gut microbes to influence the host immune system [37]. The microbiota in the digestive tract controls neutrophil migration and activity [38]. Conversely, it has the potential to induce differentiation of T cells into helper cells (such as Th1, Th2, and Th17) as well as regulatory T cells [39]. Studies have demonstrated the effectiveness of individual vitamins or their combinations with other vitamins and minerals in reducing the transmission of infectious diseases [40]. Each vitamin, either individually or in combination, is essential for initiating specific stages of the immune response. For instance, vitamins A, D, E, and C play a crucial role in modulating the immune system [41,42].

B-COMPLEX VITAMINS

The eight vitamins that make up the B-complex family of water-soluble nutrients are B1 (thiamin), B2 (riboflavin), B3 (niacin), B6 (pyridoxine), B12, folate (folic acid), biotin, and pantothenic acid. Through the release of energy from food, vitamin B1 is essential for muscular contraction and the transmission of nerve messages. It has been demonstrated that thiamin deficiency may cause severe symptoms by causing an increase in inflammation and abnormal antibody responses. Furthermore, an adequate amount of thiamine has the potential to prevent hypoxia in patients and reduce hospitalization rates [43]. This effect can be attributed to its inhibition of carbonic anhydrase isoenzymes [44].

The energy metabolism of bodily cells is regulated by vitamin B2 (riboflavin) [45]. In a study by Keil et al., vitamin B2 and UV radiation reduced the titer of MERS-CoV in products made from human plasma, but it is a major contributing factor to the occurrence of severe hypoxemia [46]. By reducing the accumulation of neutrophils in the lungs following ventilator-induced lung damage, vitamin B3, also known as niacin, exerts a potent anti-inflammatory effect. Niacin, as a constituent of NAD and NADP, has the potential to diminish the production of pro-inflammatory cytokines like IL-1, IL-6, and TNF-, thereby aiding in the regulation of cytokine storm [47,48]. Conversely, a deficiency in cobalamin, or vitamin B12, can hinder the immune system's ability to generate antibodies against viral infections. Furthermore, severe deficiency can lead to hyperhomocysteinemia, resulting in the formation of life-threatening blood clots in the brain, lungs, and lower extremities.

The majority of animal products, including meat, milk, eggs, and fish, contain vitamin B12. Additionally, nutritional yeasts include vitamin B12 and are advantageous for vegans who are at a high risk of B12 deficiency. The production of advantageous immunological responses, energy metabolism, enhancement of respiratory function, and decrease in pro-inflammatory cytokine titer are all fundamentally influenced by various vitamin B subtypes [49].

VITAMIN C

Due to its ability to donate electrons, vitamin C, also known as ascorbic acid, is an important micronutrient with pleiotropic effects on the body. It serves as a cofactor for a variety of enzymes that are essential for biosynthesis and gene regulation. The innate and adaptive immune systems' cell-mediated immunity is significantly influenced by vitamin C. In addition, vitamin C acts as an antioxidant to prevent lipid peroxidation, protein alkylation, and reactive oxygen species (ROS) from causing cellular damage due to oxidative stress [50].

The crucial role of vitamin C in infectious disorders can be attributed to its impact on various aspects: (i) facilitating the migration of neutrophils toward the site of infection, (ii) enhancing phagocytosis and the generation of oxidants, and (iii) eradicating pathogenic microorganisms [51]. By promoting neutrophil apoptosis, enhancing macrophage clearance, and reducing neutrophil necrosis, vitamin C offers protection against severe tissue damage caused by infectious pathogens.

Furthermore, studies have demonstrated that administering high doses of vitamin C to patients with sepsis and acute respiratory distress syndrome does not result in significant changes in inflammation markers, but it does improve organ dysfunction scores. Early intravenous administration of vitamin C, in combination with corticosteroids and thiamine, has also been shown to significantly reduce mortality rates in patients with severe sepsis and septic shock, while preventing the progression of organ dysfunction [52].

Nevertheless, conflicting findings exist regarding this matter. As an example, a cohort study involving 94 patients with severe sepsis and septic shock employed a triple therapy approach comprising hydrocortisone, vitamin C, and thiamine. However, this study did not yield positive outcomes concerning hospital mortality, length of hospital stay, ICU mortality, ICU length of stay, and the time required for vasopressor independence [53].

METABOLIC ROLE OF WATER-SOLUBLE VITAMINS AND EFFECT OF METABOLIC PERTURBATIONS IN VIRAL INFECTIONS

B6 vitamin toxicity can result in painful skin sores, light sensitivity, and nerve damage [54]. Niacin, a B3 vitamin, comes in large doses that can cause liver damage, high blood sugar, vomiting, and skin rashes [55].

Since vitamin C is water-soluble and frequently eliminated by the body, toxicity ordinarily does not happen. The presence of extra ascorbic acid in the urine causes a false-positive result for sugar. Copper absorption is hampered by high vitamin C levels [56]. People who have kidney stones should stay away from vitamin C since it can turn into oxalate [57]. However, some studies indicate that this change of vitamin C in urine only occurs after the urine has left the body [58].

NUTRITIONAL SUPPLEMENTATION OF VITAMINS B AND C AS VIRAL PROPHYLACTIC

Cellular DNA is shielded by vitamin C from destruction by free radicals and mutagens. It guards against damaging genetic changes within cells and shields lymphocytes from chromosome mutations [59]. According to Kronhausen et al. [63], vitamin C helps to shield the central nervous system from free radical damage and prevents damage to the lungs from occurring. Ascorbic acid pre-treatment effectively reduced the acute lung damage brought on by the introduction of superoxide anion-free oxygen radicals to the trachea in a study of guinea pigs [60]. Additionally, ascorbic acid has been studied in mice as an anti-inflammatory antioxidant. High doses administered following the injury but not beforehand were successful in suppressing edema [61].

Vitamin C serves as an antioxidant that can regenerate vitamin E, thereby indirectly assisting in the mitigation of lipid damage caused by free radicals. Consequently, it is not surprising that these two nutrients can effectively collaborate to reduce the detrimental process of lipid peroxidation. This observed decline was evident in human and animal studies involving individuals with diabetes, cerebral arteriosclerosis, or a heart condition [62]. The combined intake of vitamins C and E can also help decrease the risk of blood clot formation, which is associated with an increased likelihood of stroke [63]. Additionally, the inclusion of vitamin A may further enhance the beneficial effects derived from the synergistic combination of vitamins C and E. Using this combination, researchers were able to improve the "characteristics of enzymatic and non-enzymatic antioxidant protection of the liver" in mice in one experiment [64].

THERAPEUTIC BENEFITS OF VITAMINS B AND C IN EBOLA: EXAMPLES FROM CLINICAL REPORTS

Current international guidelines recommend the use of micronutrient supplementation in cases of micronutrient deficiencies [65], and these guidelines are often followed in supportive care for EVD [66]. This may involve the administration of multivitamins, which have shown clinical benefits in the treatment of various infectious diseases such as HIV, dengue fever, and tuberculosis [67]. Additionally, multivitamins provide essential cofactors that have demonstrated efficacy in the management of sepsis. Studies on vitamin supplementation have also highlighted the significant advantages for populations with poor baseline nutritional status, particularly those in sub-Saharan Africa [68]. While data on the effects of multivitamin supplementation on patient-centered outcomes in EVD are currently unavailable, previous research suggests that the use of affordable multivitamins may offer clinical benefits to patients with EVD.

During the EVD epidemic from 2014 to 2016, the International Medical Corps (IMC) established and supervised five Ebola treatment units (ETUs) in Liberia and Sierra Leone. These ETUs provided medical care to a total of 478 individuals diagnosed with the disease. Comprehensive clinical, epidemiological, and laboratory data were collected for the majority of patients, resulting in a significant global database that could be utilized for studying the impact of different management strategies on patient outcomes. In this particular study, the focus was on evaluating the effect of early multivitamin supplementation on mortality rates among EVD patients admitted to the five IMC ETUs. Out of the 313 patients (73.8%) who received multivitamin treatment during their time in the ETUs, 261 (61.6%) patients were administered daily multivitamins within the first 48 hours of their treatment [69].

The average duration of multivitamin treatment for all cases was 5 days (interquartile range: 3, 10). The frequency distributions depicting the duration and timing of multivitamin therapy are presented. Patients who received multivitamins during the initial 48 hours exhibited significantly lower rates of anorexia, bleeding, confusion, and fever while under ETU care. Additionally, this group was more likely to be female and have low CT values. Within the first 48 hours of care, vitamin C or vitamin A co-treatment was administered to 248 patients (58.9%) and 230 patients (54.2%), respectively. Unfortunately, out of the total number of patients, 244 (57.5%) passed away while receiving care in an ETU [70].

Consistent with previous reports [71], there were no significant differences in mortality rates among the five ETU sites. A notable association was found between viral load at admission and mortality: 44.7% of patients who died had CT values below 22, whereas only 16.8% had CT values above 22 ($p < 0.001$). Patients who passed away while receiving care in an ETU had significantly higher rates of diarrhea, bleeding, dyspnea, and dysphagia. The observed mortality rate for patients treated with multivitamins within 48 hours of admission was 53.6%, while it was 63.8% for patients who did not receive treatment within this timeframe. Among patients who passed away within 48 hours of ETU admission, the median survival time was 4 days (interquartile range: 2, 6) for those without multivitamin treatment and 5 days (interquartile range: 4, 7) for those who received multivitamins.

CHALLENGES AND DOSE-RELATED ILL EFFECTS OF VITAMINS B AND C IN EBOLA

In vitro studies and a mouse model have shown the effectiveness of the anti-Ebola medications clomiphene and toremifene. To assess their efficacy, four groups were exposed to the Ebola virus: the clomiphene group ($n = 10$) and its control group ($n = 10$), and the toremifene group ($n = 10$) and its control group ($n = 7$). Following vaccination, both treatment groups received the medication at a dose of 60 mg/kg on days 0 and 1, followed by administration every other day for a total of six doses. In the clomiphene group, 90% of patients survived (in contrast to 0% in the control group, $p < 0.0001$), while in the toremifene group, 50% of patients survived (compared to 0% in the control group, $p=0.0441$) [72]. Another promising finding is related to amiodarone, which showed significant potential in inhibiting Ebola virus cell entry in a preliminary screening study, even at concentrations typically observed with standard amiodarone dosing [73]. However, it's important to note that this screening study was conducted solely in cell cultures, and no animal models were involved. Amiodarone demonstrated an impact on various cell types, including macrophages, monocytes, and erythrocytes, but not on human hepatocytes. The management and treatment of the Ebola virus epidemic have been fraught with numerous challenges. In Africa, the reported case mortality rate for this outbreak stands at 70%. Additionally, during the only filovirus outbreak in Western Europe, which took place in Germany and Yugoslavia in 1967, 22% of individuals unfortunately lost their lives [74]. Key aspects of treatment include fluid resuscitation, correction of electrolyte abnormalities, prevention and treatment of concurrent and superinfections, and the prevention of complications associated with shock. Various potential therapies are currently under development or already in use to address and prevent the Ebola virus during this ongoing epidemic.

CONCLUSION

The Ebola virus possesses the potential to be employed as a biological weapon against adversaries. However, its effectiveness for mass destruction is hindered by its inability to thrive in open air [76]. An instance highlighting its use as a bio-weapon is when North Korean state media, as reported by the British broadcast channel (BBC) in 2015, suggested that the US military had developed the disease as a biological weapon [75]. This revelation should serve as a wake-up call for the international community, urging them to cooperate and take action. Previous outbreaks have demonstrated that, given its exponential growth rate and the absence of immediate care, the cumulative number of Ebola infections could reach unprecedented levels in the coming years. Currently, patients are being sent home without any cure, leaving little opportunity for control efforts. The highly lethal and contagious nature of this virus underscores the critical need for dependable diagnostic methods [76]. It is imperative that we demonstrate unwavering commitment in responding to this natural disaster and mitigating its present and future impact.

REFERENCES

1. Taylor, L. H., Latham, S. M., & Woolhouse, M. E. (2001). Risk factors for human disease emergence. *Philosophical Transactions of the Royal Society of London. Series B: Biological Sciences, 356*(1411), 983–989.
2. Rasheed, A., Ullah, S., Naeem, S., Zubair, M., Ahmad, W., & Hussain, Z. (2014). Occurrence of HCV genotypes in different age groups of patients from Lahore, Pakistan. *Advancements in Life Sciences, 1*(2), 89–95.
3. Lennemann, N. J., Rhein, B. A., Ndungo, E., Chandran, K., Qiu, X., & Maury, W. (2014). Comprehensive functional analysis of N-linked glycans on Ebola virus GP1. *MBio, 5*(1), e00862-13.
4. Drosten, C., Kümmerer, B. M., Schmitz, H., & Günther, S. (2003). Molecular diagnostics of viral hemorrhagic fevers. *Antiviral Research, 57*(1–2), 61–87.
5. Yaqoob, A., Shehzad, U., Ahmad, Z., Naseer, N., & Bashir, S. (2015). Effective treatment strategies against Ebola virus. *Advancements in Life Sciences, 2*(4), 176–182.
6. Deng, I. M., Duku, O., Gillo, A. L., Idris, A. A., Lolik, P., el Tahir, B.,... & WHO. (1978). Ebola hemorrhagic-fever in Sudan, 1976-report of a who international study team. *Bulletin of the World Health Organization, 56*(2), 247–270.
7. Elliott, L. H., Sanchez, A., Holloway, B. P., Kiley, M. P., & McCormick, J. B. (1993). Ebola protein analyses for the determination of genetic organization. *Archives of Virology, 133*, 423–436.
8. Miranda, M. E., Ksiazek, T. G., Retuya, T. J., Khan, A. S., Sanchez, A., Fulhorst, C. F.,... & Peters, C. J. (1999). Epidemiology of Ebola (subtype Reston) virus in the Philippines, 1996. *The Journal of Infectious Diseases, 179*(suppl_1), S115–S119.

9. Miranda, M. E. G., & Miranda, N. L. J. (2011). Reston ebolavirus in humans and animals in the Philippines: a review. *The Journal of Infectious Diseases, 204*(suppl_3), S757–S760.

10. Marí Saéz, A., Weiss, S., Nowak, K., Lapeyre, V., Zimmermann, F., Düx, A.,... & Leendertz, F. H. (2015). Investigating the zoonotic origin of the West African Ebola epidemic. *EMBO Molecular Medicine, 7*(1), 17–23.

11. Bausch, D. G., Towner, J. S., Dowell, S. F., Kaducu, F., Lukwiya, M., Sanchez, A.,... & Rollin, P. E. (2007). Assessment of the risk of Ebola virus transmission from bodily fluids and fomites. *The Journal of Infectious Diseases, 196*(suppl_2), S142–S147.

12. Pinzon, J. E., Wilson, J. M., Tucker, C. J., Arthur, R., Jahrling, P. B., & Formenty, P. (2004). Trigger events: enviroclimatic coupling of Ebola hemorrhagic fever outbreaks. *The American Journal of Tropical Medicine and Hygiene, 71*(5), 664–674.

13. Kobinger, G. P., Leung, A., Neufeld, J., Richardson, J. S., Falzarano, D., Smith, G.,... & Weingartl, H. M. (2011). Replication, pathogenicity, shedding, and transmission of Zaire ebolavirus in pigs. *Journal of Infectious Diseases, 204*(2), 200–208.

14. Carette, J. E., Raaben, M., Wong, A. C., Herbert, A. S., Obernosterer, G., Mulherkar, N.,... & Brummelkamp, T. R. (2011). Ebola virus entry requires the cholesterol transporter Niemann-Pick C1. *Nature, 477*(7364), 340–343.

15. Miller, E. H., Obernosterer, G., Raaben, M., Herbert, A. S., Deffieu, M. S., Krishnan, A.,... & Chandran, K. (2012). Ebola virus entry requires the host-programmed recognition of an intracellular receptor. The EMBO Journal, 31(8), 1947–1960.

16. Kondratowicz, A. S., Lennemann, N. J., Sinn, P. L., Davey, R. A., Hunt, C. L., Moller-Tank, S.,... & Maury, W. (2011). T-cell immunoglobulin and mucin domain 1 (TIM-1) is a receptor for Zaire Ebolavirus and Lake Victoria Marburgvirus. *Proceedings of the National Academy of Sciences, 108*(20), 8426–8431.

17. Shimojima, M., Takada, A., Ebihara, H., Neumann, G., Fujioka, K., Irimura, T.,... & Kawaoka, Y. (2006). Tyro3 family-mediated cell entry of Ebola and Marburg viruses. *Journal of Virology, 80*(20), 10109–10116.

18. Schornberg, K., Matsuyama, S., Kabsch, K., Delos, S., Bouton, A., & White, J. (2006). Role of endosomal cathepsins in entry mediated by the Ebola virus glycoprotein. *Journal of Virology, 80*(8), 4174–4178.

19. Takahashi, K., Halfmann, P., Oyama, M., Kozuka-Hata, H., Noda, T., & Kawaoka, Y. (2013). DNA topoisomerase 1 facilitates the transcription and replication of the Ebola virus genome. *Journal of Virology, 87*(16), 8862–8869.

20. Wool-Lewis, R. J., & Bates, P. (1998). Characterization of Ebola virus entry by using pseudotyped viruses: identification of receptor-deficient cell lines. *Journal of Virology, 72*(4), 3155–3160.

21. Feldmann, H., Nichol, S. T., Klenk, H. D., Peters, C. J., & Sanchez, A. (1994). Characterization of filoviruses based on differences in structure and antigenicity of the virion glycoprotein. *Virology, 199*(2), 469–473.

22. Volchkov, V. E., Feldmann, H., Volchkova, V. A., & Klenk, H. D. (1998). Processing of the Ebola virus glycoprotein by the proprotein convertase furin. *Proceedings of the National Academy of Sciences, 95*(10), 5762–5767.

23. Kaletsky, R. L., Simmons, G., & Bates, P. (2007). Proteolysis of the Ebola virus glycoproteins enhances virus binding and infectivity. *Journal of Virology, 81*(24), 13378–13384.

24. Hood, C. L., Abraham, J., Boyington, J. C., Leung, K., Kwong, P. D., & Nabel, G. J. (2010). Biochemical and structural characterization of cathepsin L-processed Ebola virus glycoprotein: implications for viral entry and immunogenicity. *Journal of Virology, 84*(6), 2972–2982.

25. Lee, J. E., Fusco, M. L., Hessell, A. J., Oswald, W. B., Burton, D. R., & Saphire, E. O. (2008). Structure of the Ebola virus glycoprotein bound to an antibody from a human survivor. *Nature, 454*(7201), 177–182.

26. Brecher, M., Schornberg, K. L., Delos, S. E., Fusco, M. L., Saphire, E. O., & White, J. M. (2012). Cathepsin cleavage potentiates the Ebola virus glycoprotein to undergo a subsequent fusion-relevant conformational change. *Journal of Virology, 86*(1), 364–372.

27. Jeffers, S. A., Sanders, D. A., & Sanchez, A. (2002). Covalent modifications of the ebola virus glycoprotein. *Journal of Virology, 76*(24), 12463–12472.

28. Brindley, M. A., Hughes, L., Ruiz, A., McCray Jr, P. B., Sanchez, A., Sanders, D. A., & Maury, W. (2007). Ebola virus glycoprotein 1: identification of residues important for binding and postbinding events. *Journal of Virology, 81*(14), 7702–7709.

29. Weissenhorn, W., Carfí, A., Lee, K. H., Skehel, J. J., & Wiley, D. C. (1998). Crystal structure of the Ebola virus membrane fusion subunit, GP2, from the envelope glycoprotein ectodomain. *Molecular Cell, 2*(5), 605–616.

30. Yang, Z. Y., Duckers, H. J., Sullivan, N. J., Sanchez, A., Nabel, E. G., & Nabel, G. J. (2000). Identification of the Ebola virus glycoprotein as the main viral determinant of vascular cell cytotoxicity and injury. *Nature Medicine, 6*(8), 886–889.

31. Kondratowicz, A. S., Hunt, C. L., Davey, R. A., Cherry, S., & Maury, W. J. (2013). AMP-activated protein kinase is required for the macropinocytic internalization of ebolavirus. *Journal of Virology, 87*(2), 746–755.

32. Chandran, K., Sullivan, N. J., Felbor, U., Whelan, S. P., & Cunningham, J. M. (2005). Endosomal proteolysis of the Ebola virus glycoprotein is necessary for infection. *Science, 308*(5728), 1643–1645.

33. Gombart, A. F., Pierre, A., & Maggini, S. (2020). A review of micronutrients and the immune system-working in harmony to reduce the risk of infection. *Nutrients, 12*(1), 236.

34. Maggini, S., Beveridge, S., Sorbara, P. J., & Senatore, G. (2009). Feeding the immune system: the role of micronutrients in restoring resistance to infections. *CABI Reviews,* (2008), 1–21.

35. Ooi, J. H., Li, Y., Rogers, C. J., & Cantorna, M. T. (2013). Vitamin D regulates the gut microbiome and protects mice from dextran sodium sulfate-induced colitis. *The Journal of Nutrition, 143*(10), 1679–1686.

36. Cantorna, M. T., McDaniel, K., Bora, S., Chen, J., & James, J. (2014). Vitamin D, immune regulation, the microbiota, and inflammatory bowel disease. *Experimental Biology and Medicine, 239*(11), 1524–1530.

37. Wu, H. J., & Wu, E. (2012). The role of gut microbiota in immune homeostasis and autoimmunity. *Gut Microbes, 3*(1), 4–14.

38. Owaga, E., Hsieh, R. H., Mugendi, B., Masuku, S., Shih, C. K., & Chang, J. S. (2015). Th17 cells as potential probiotic therapeutic targets in inflammatory bowel diseases. *International Journal of Molecular Sciences, 16*(9), 20841–20858.

39. Francino, M. P. (2014). Early development of the gut microbiota and immune health. *Pathogens, 3*(3), 769–790.

40. Gariballa, S. (2005). Vitamin and mineral supplements for preventing infections in older people. *BMJ, 331*(7512), 304–305.

41. Galanakis, C. M. (2020). The food systems in the era of the coronavirus (COVID-19) pandemic crisis. *Foods, 9*(4), 523.

42. Thirumdas, R., Kothakota, A., Pandiselvam, R., Bahrami, A., & Barba, F. J. (2021). Role of food nutrients and supplementation in fighting against viral infections and boosting immunity: a review. *Trends in Food Science & Technology, 110*, 66–77.

43. Shakoor, H., Feehan, J., Mikkelsen, K., Al Dhaheri, A. S., Ali, H. I., Platat, C.,... & Apostolopoulos, V. (2021). Be well: A potential role for vitamin B in COVID-19. *Maturitas, 144*, 108–111.

44. Özdemir, Z. Ö., Şentürk, M., & Ekinci, D. (2013). Inhibition of mammalian carbonic anhydrase isoforms I, II and VI with thiamine and thiamine-like molecules. *Journal of Enzyme Inhibition and Medicinal Chemistry, 28*(2), 316–319.

45. Powers, H. J. (2003). Riboflavin (vitamin B-2) and health. *The American Journal of Clinical Nutrition, 77*, 1352–1360.

46. Jones, H. D., Yoo, J., Crother, T. R., Kyme, P., Ben-Shlomo, A., Khalafi, R.,... & Shimada, K. (2015). Nicotinamide exacerbates hypoxemia in ventilator-induced lung injury independent of neutrophil infiltration. *PLOS ONE, 10*(4), e0123460.

47. Mikkelsen, K., Stojanovska, L., Prakash, M., & Apostolopoulos, V. (2017). The effects of vitamin B on the immune/cytokine network and their involvement in depression. *Maturitas, 96*, 58–71.

48. Liu, B., Li, M., Zhou, Z., Guan, X., & Xiang, Y. (2020). Can we use interleukin-6 (IL-6) blockade for coronavirus disease 2019 (COVID-19)-induced cytokine release syndrome (CRS)? *Journal of Autoimmunity, 111*, 102452.

49. Mikkelsen, K., & Apostolopoulos, V. (2019). Vitamin B1, B2, B3, B5, and B6 and the immune system. In Mahmoudi, M., & Rezaei, N. (eds.), *Nutrition and Immunity*, 115–125.Cham: Springer. https://doi.org/10.1007/978-3-030-16073-9_7

50. Traber, M. G., & Stevens, J. F. (2011). Vitamins C and E: beneficial effects from a mechanistic perspective. *Free Radical Biology and Medicine, 51*(5), 1000–1013.

51. Chuangchot, N., Boonthongkaew, C., Phoksawat, W., Jumnainsong, A., Leelayuwat, C., & Leelayuwat, N. (2020). Oral vitamin C treatment increases polymorphonuclear cell functions in type 2 diabetes mellitus patients with poor glycemic control. *Nutrition Research, 79,* 50–59.

52. Hager, D. N., Hooper, M. H., Bernard, G. R., Busse, L. W., Ely, E., Gaieski, D. F.,... & Martin, G. S. (2019). The Vitamin C, Thiamine and Steroids in Sepsis (VICTAS) Protocol: a prospective, multi-center, double-blind, adaptive sample size, randomized, placebo-controlled, clinical trial. *Trials, 20*(1), 1–16.

53. Litwak, J. J., Cho, N., Nguyen, H. B., Moussavi, K., & Bushell, T. (2019). Vitamin C, hydrocortisone, and thiamine for the treatment of severe sepsis and septic shock: a retrospective analysis of real-world application. *Journal of Clinical Medicine, 8*(4), 478.

54. Vrolijk, M. F., Opperhuizen, A., Jansen, E. H., Hageman, G. J., Bast, A., & Haenen, G. R. (2017). The vitamin B6 paradox: supplementation with high concentrations of pyridoxine leads to decreased vitamin B6 function. *Toxicology in Vitro, 44,* 206–212.

55. Ellsworth, M. A., Anderson, K. R., Hall, D. J., Freese, D. K., & Lloyd, R. M. (2014). Acute liver failure secondary to niacin toxicity. *Case Reports in Pediatrics, 2014,* 692530–692533.

56. Finley, E. B., & Cerklewski, F. L. (1983). Influence of ascorbic acid supplementation on copper status in young adult men. The American Journal of Clinical Nutrition, *37*(4), 553–556.

57. Piesse, J. W. (1985). Nutritional factors in calcium containing kidney stones with particular emphasis on vitamin C. *International Clinical Nutrition Review, 5*(110), 29.

58. Wandzilak, T. R., D'andre, S. D., Davis, P. A., & Williams, H. E. (1994). Effect of high dose vitamin C on urinary oxalate levels. *The Journal of Urology, 151*(4), 834–837.

59. Gaby, S. K., Bendich, A., & Singh, V. N. (1991). *Vitamin Intake and Health: A Scientific Review.* Marcel Dedder, New York.

60. Becher, G., & Winsel, K. (1989). Short scientific report. Vitamin C lessens superoxide anion (O2)-induced bronchial constriction. *Zeitschrift fur Erkrankungen der Atmungsorgane, 173*(1), 100–104.

61. Spillert, C. R., Spillert, K. R., Hollinshead, M. B., & Lazaro, E. J. (1989). Inhibitory effect of high dose ascorbic acid on inflammatory edema. *Agents and Actions, 27,* 401–402.

62. Karagezian, K. G., & Gevorkian, D. M. (1989). Phospholipid-glycerides, cross-resistance of erythrocytes, malonic dialdehyde level and alpha-tocopherol levels in the plasma and erythrocytes of rats with alloxan diabetes before and after combined antioxidant therapy. *Voprosy meditsinskoi khimii, 35*(5), 27–30.

63. Kronhausen, E., Kronhausen, P., & Demopoulos, H. B. (1989) *Formula for Life.* William Morrow and Co., New York, 102.

64. Kuvshinnikov, V. A., Morozkina, T. S., Svirnovskiĭ, A. I., Poliakova, Z. I., Stre'lnikov, A. V., Oletskiĭ, E. I.,... & Sukolinskiĭ, V. N. (1989). Use of the antioxidant complex of vitamins A, E and C in murine leukemia. *Gematologiia i Transfuziologiia, 34*(8), 23–28.

65. World Health Organization. (2014). *Clinical management of patients with viral haemorrhagic fever: a pocket guide for the front-line health worker: interim emergency guidance-generic draft for West African adaptation 30 March 2014* (No. WHO/HSE/PED/AIP/14.05). World Health Organization.

66. Barry, M., Traoré, F. A., Sako, F. B., Kpamy, D. O., Bah, E. I., Poncin, M.,... & Touré, A. (2014). Ebola outbreak in Conakry, Guinea: epidemiological, clinical, and outcome features. *Medecine et maladies infectieuses, 44*(11–12), 491–494.

67. Ahmed, S., Finkelstein, J. L., Stewart, A. M., Kenneth, J., Polhemus, M. E., Endy, T. P.,... & Mehta, S. (2014). Micronutrients and dengue. *The American Journal of Tropical Medicine and Hygiene, 91*(5), 1049.

68. Carr, A. C., Shaw, G. M., & Natarajan, R. (2015). Ascorbate-dependent vasopressor synthesis: a rationale for vitamin C administration in severe sepsis and septic shock?. *Critical Care, 19*(1), 1–8.

69. Martineau, A. R., Jolliffe, D. A., Hooper, R. L., Greenberg, L., Aloia, J. F., Bergman, P.,... & Camargo, C. A. (2017). Vitamin D supplementation to prevent acute respiratory tract infections: systematic review and meta-analysis of individual participant data. *bmj, 356*, i6583.

70. Skrable, K., Roshania, R., Mallow, M., Wolfman, V., Siakor, M., & Levine, A. C. (2017). The natural history of acute Ebola Virus Disease among patients managed in five Ebola treatment units in West Africa: a retrospective cohort study. *PLOS Neglected Tropical Diseases, 11*(7), e0005700.

71. Marik, P. E., Khangoora, V., Rivera, R., Hooper, M. H., & Catravas, J. (2017). Hydrocortisone, vitamin C, and thiamine for the treatment of severe sepsis and septic shock: a retrospective before-after study. *Chest, 151*(6), 1229–1238.

72. Schepers, A. G., Snippert, H. J., Stange, D. E., van den Born, M., van Es, J. H., van de Wetering, M., & Clevers, H. (2012). Lineage tracing reveals Lgr5+ stem cell activity in mouse intestinal adenomas. *Science, 337*(6095), 730–735.

73. Gehring, G., Rohrmann, K., Atenchong, N., Mittler, E., Becker, S., Dahlmann, F.,... & von Hahn, T. (2014). The clinically approved drugs amiodarone, dronedarone and verapamil inhibit filovirus cell entry. *Journal of Antimicrobial Chemotherapy, 69*(8), 2123–2131.

74. Feldmann, H., & Geisbert, T. W. (2011). Ebola haemorrhagic fever. *The Lancet, 377*(9768), 849–862.

75. Dixon, M. G., & Schafer, I. J. (2014). Ebola viral disease outbreak-West Africa, 2014. *Morbidity and Mortality Weekly Report, 63*(25), 548.

76. Salvaggio, M. R., & Baddley, J. W. (2004). Other viral bioweapons: Ebola and Marburg hemorrhagic fever. *Dermatologic clinics, 22*(3), 291–302.

10

Novel vitamin interventional therapies and approaches against infectious viral diseases

BHAVYA JAIN, NIDA SIDDIQUI, AND LOKESH KORI

INTRODUCTION

Vitamins and minerals are organic compounds that are required by our body in an adequate amount for various metabolic processes. Basically, they keep us healthy and help our body in fighting against infection, wound healing, and tissue repair. We get vitamins and minerals from the food we eat. A proper balanced diet that contains all the nutrients in the right amount is the basic requirement to live a healthy and disease-free life. There are plenty of sources to get vitamins and minerals. It is best to get vitamins and minerals by eating a variety of healthy unrefined foods. Vitamins and minerals are called micronutrients as they are required in minute quantity. Vitamins play an important role in combating various viral diseases. In general, both water-soluble and hydrophobic vitamins are known to have diverse biological functions. Vitamin A is fat-soluble vitamin, basically retinols and retinyl esters. It plays major role in providing immunity, cellular-repair, organs-functioning, cell growth, and differentiation. It is an essential component in rhodopsin signaling, and it's a light-sensitive protein in the retina that responds to the light entering the eyes and works in vision. Various clinical research studies found that vitamin A has a great role as a therapeutic against viral diseases like COVID-19, influenza, and human immunodeficiency virus (HIV). Studies show that it has a positive effect on COVID-19 infection. SARS-CoV-2 infects respiratory epithelial cells by its cellular receptor angiotensin-converting enzyme 2 (ACE-2), which results in viral pneumonia with severe inflammation that results in deep damage and hampers the functioning of the lungs and other organs. The symptoms and severity vary from person to person depending upon the exposure and co-morbidities, all sum up to, affecting immune system and delaying the process of recovery. Serious infection results in nutritional deficiencies, and this leads to more infectious disease and impaired recovery. In the case of viral infection measles, associated vitamin A increases the risk of infection, and on the other hand, appropriate supplementation of vitamin A lowers the mortality and positively regulates recovery. Vitamin A also plays same role in COVID-19-infected individuals as it regulates both innate immunity and adaptive immunity by promoting clearance of primary infection and reducing the risk of secondary infection. It protects the inner lining of respiratory tract, lowers the damage caused by the inflammation, and promotes recovery of the tract. Supplementation of vitamin A may counteract adverse effects of SARS-CoV2 on the angiotensin system along with minimizing the adverse effects of COVID-19 therapies on the individuals [1].

Vitamin B ranges from B1 to B7 and B12 and folic acid. These vitamins help in getting energy from the food by processing it. They play a vital role in generation of red blood cells. We can get vitamins from proteins including fish, poultry, meat, eggs, dairy products, and leafy green vegetables. If not taken in an adequate amount, it gives rise to diseases like anemia (lack of vitamins B12 and B6). There are various therapeutic approaches by vitamin B against viral diseases. A meta-analysis shows that vitamin B is an adjunct therapy for viral infection in the treatment of prolonged symptoms of COVID-19 and focuses on the muscle–gut–brain axis [2]. Through meta-analysis, it is clear that vitamin B12 in the form of methylcobalamin and cyanocobalamin may increase serum vitamin B12 level which resulted in decreased serum methylmalonic acid and homocysteine concentration, thus decreasing pain intensity, memory loss, and impaired concentrations [2].

Vitamin C is a potent antioxidant and a cofactor for a family of biosynthetic and gene regulatory enzymes. Vitamin C helps in different immune defenses by supporting various cellular functions of both the innate and adaptive immune system. Vitamin C supports epithelial barrier function against pathogens and promotes the

DOI: 10.1201/9781003435686-10

oxidant scavenging activity of the skin, thereby potentially protecting against environmental oxidative stress. Vitamin C accumulates in phagocytic cells, such as neutrophils, and can enhance chemotaxis, phagocytosis, generation of reactive oxygen species, and ultimately microbial killing. The most extensively studied virus in human infection is the common cold caused by rhinoviruses. An experiment was performed between the groups of people taking vitamin C during cold and those who avoid the intake of it. Although vitamin C intake does not lower the cases of cold in the general population, it halved the number of colds in physically active people. Continuous intake of vitamin C has reduced the duration of colds, indicating a biological effect in an experiment. Two controlled trials revealed a statistically significant dose–response, for the incubation period and duration of common cold symptoms persisting, with up to 6–8 g/day of vitamin C. Negative findings of some therapeutic common cold studies might be due to low amount of doses 3–4 g/day of vitamin C. It clearly signifies that a higher amount of vitamin C administration is required to show an impactful result. Various positive outcomes were observed in three controlled trials that found that vitamin C worked against pneumonia. Two controlled trials revealed treatment benefits of vitamin C in pneumonia patients. One controlled trial reported treatment benefits in tetanus patients [3]. Vitamin C acts as an antiviral agent since it increases the survival rate in COVID-19-infected individuals by attenuating the enhancement of antiviral cytokines and free radical formation which ultimately hampers the viral load [4].

Vitamin D deficiency leads to an increased risk of metabolic disorders such as obesity and diabetes, cancer, tuberculosis (TB), and hepatitis C. Vitamin D helps in maintenance of bone strength and muscle, induces lipolysis, decreases lipogenesis, shows anti-inflammatory, proapoptotic, and antiangiogenic properties, and also interacts with interferon alfa which inhibits Hepatitis B Virus (HCV) replication. HIV-infected persons are usually vitamin D deficient as certain antiretroviral drugs such as efavirenz are known to interfere with vitamin D metabolism, where efavirenz induces 24-hydroxylase, a cytochrome P450 (CYP450) enzyme which inactivates 25(OH)D and 1,25(OH)$_2$D [5]. Recent analysis showed that patients with low vitamin D levels had higher risk for progression to AIDS and increased non-AIDS-related disorder susceptibility further leading to death [6]. Also, low vitamin D levels in HIV-infected pregnant women increase the risk of adverse birth and infant health [7].

Vitamin E (VE) (α-tocopherol) is a lipid-soluble vitamin having antioxidant, anti-inflammatory, gene regulatory, and immune-modulatory properties [8,9]. VE deficiency is measured by levels of α-tocopherol less than 0.50 mg/dL [10]. VE supplementation in hepatitis B virus (HBV)-infected patients has shown promising results by clearing hepatitis B antigen and antibody production against it as well as HBV-DNA loss. VE acts through regulating expression of critical genes at post-transcriptional level by modulating miRNA exhibiting antiviral activity. As miRNA plays a key role in developing viral infections, VE modulates these miRNAs to suppress HBV infection. VE supplements influence the hepatic expression of a total of eight miRNAs (i.e. miRNA-16, miRNA-21, miRNA-122, miRNA-125b, miRNA-146a, miRNA-155, miRNA-181a, and miRNA-223). VE increases the expression of miRNAs such as miRNA-125b and miRNA-181a which further regulate innate and adaptive immunity, miRNA-146 which decreases the production of antiviral cytokines and modulates inflammatory response, and miRNA-155 which downregulates HBV replication [11].

IMMUNOMODULATORY APPROACHES OF VITAMIN D AND ITS NOVEL OUTCOMES

In clinical experiments, performed on influenza model of mice using intranasal inoculation of influenza viruses in Gulo(-/-)**mice,** which are unable to synthesize vitamin, it was evident that the viral load increased in lungs but production of interferon (IFN)α/β and antiviral cytokines decreased. Contrary to this, invasion of inflammatory cells in the lungs, pro-inflammatory cytokines, TNF-α, and interleukin α/β was seen. From this study, it is clear that vitamin C creates antiviral responses by increased production of IFN-α/β [12]. Vitamin D and its receptors are key regulators of immune functions such as antibody production, lymphocyte proliferation, and modulation of cytokine response where it upregulates IL-4, IL-10, TNF-α, and TGF-β and downregulates IFN-γ and IL-12, suggesting 1,25-dihydroxy vitamin D3 as hormone like immunomodulator [13–18]. Vitamin D receptor (VDR) also plays a key role in retinoic acid receptor signaling pathway by

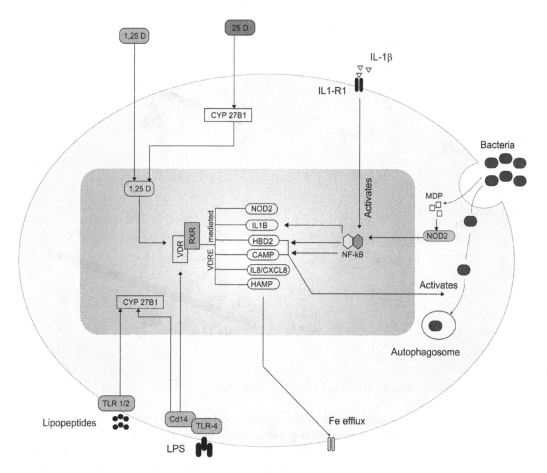

Figure 10.1 Immunomodulatory roles of vitamin D and its molecular cascades.

forming heterodimer with nuclear retinoid X receptor alpha (RXRA). Also, VDR gene polymorphism has been seen to be associated with different infection susceptibilities such as HBV, HIV-1, human T-cell lymphotropic virus type-1 (HTLV-1), measles, and rubella virus [19–22]. Due to single nucleotide polymorphism in the vitamin D receptor genes, variations in adaptive immune response have been observed in response to rubella [23,24] and measles vaccine which shows VDR's significant role in vaccine-induced immunity as well [25]. Figure 10.1 highlights the immunomodulatory effects of vitamin D and their molecular mechanism.

Role of vitamin D has also been delineated in Epstein–Barr virus (EBV) infections. Infections make one more susceptible to MS which amplifies with the age of EBV infection. Also, an inverse association between MS infection and vitamin D serum levels has been seen. High levels of vitamin D serum may lead to reduced risk of MS, and vitamin D supplementation can help prevent MS. The mechanism behind the protective role of vitamin D against MS risk is still not known, but its immunomodulatory effects might contribute to it. A significant 5%–6% increase in regulatory T cells was observed in individuals taking vitamin D supplements. MS patients taking 1000 IU/day for 6 months showed an increase in TGF-β1 levels [26]. Vitamin D doses lead to anti-inflammatory cytokine production and reduced Major Histocompatibility Complex (MHC) presentation, and decrease in Th1 and Th17 cell populations with decrease in IFN-γ and IL-17 has been seen in animal and in vitro studies [27]. However, more in vivo studies are needed to make any conclusive correlation between vitamin D and MS infection. Higher 25(OH)D levels have also seen to prevent MS infection relapses to a great extend ([8–30]. Majorly, the effect of vitamin D on MS infections is through immunomodulation affecting inflammatory response [31]. Through transcriptome database studies, nine miRNAs were found to be differentially expressed in peripheral blood between relapsing MS patients and normal groups. System biology analysis of these miRNAs showed their association with EBV and vitamin D [32] (Figure 10.2).

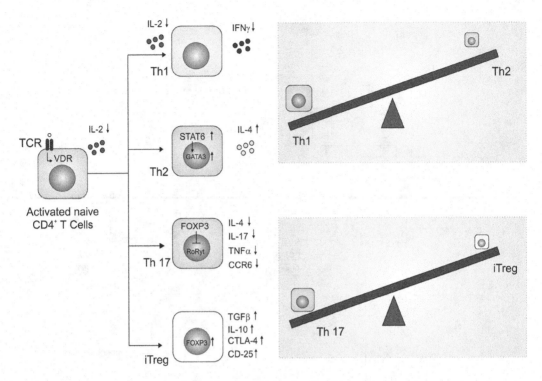

Figure 10.2 Immunomodulatory effects of vitamin D on CD4+ T cells.

Vitamin D exhibits antimycobacterial activity and bactericidal properties through increase in cathelicidin gene expression in macrophages in innate immunity, whereas in adaptive immunity it targets T cells by decreasing activation of Th1 and Th17 cells and shifts toward Th2 and Treg cells. IL-2 and IFN-γ secreted by Th1 cells which play an important role in regulation of cell-mediated immunity against viral infections are usually suppressed during retroviral infections. However, their production was significantly restored by giving vitamin E supplementation at 15-, 150-, and 450-fold, whereas IL-6 and IL-10 produced by Th2 cells, responsible for humoral immune response regulation which is increased during retroviral infection which significantly decreased by vitamin E supplementation at all levels [33]. Vitamin E also helps fight against influenza virus by increasing T helper 1 cytokine production which usually decreases with age as prostaglandin E$_2$ (PGE$_2$) production increases with age which suppresses Th1 cytokine production. It also increased IL-2 and IFN-γ production which led to lower pulmonary viral titer and decreased IL-1β and TNF-α production. It enhances immune response mainly by reducing PGE$_2$ synthesis [34]. It interacts with macrophages directly to upregulate IL-1 and IL-2. It prevents lipid peroxidation of cell membranes acting as a free radical scavenger [35].

VITAMIN E AND ANTIOXIDANT MODULATORY ROLES IN VIRAL HEPATITIS

Vitamin E plays anti-oxidative roles in patients with viral hepatitis C, and it reduces oxidative stress induced by liver damage through reduction in oxidative stress level markers such as thioredoxin (TRX) and serum alanine aminotransferase (ALT). Even increase in these markers was seen on cessation of VE supplementation [36]. It has also been observed that VE supplementation with Coenzyme Q10, selenium, and methionine leads to faster healing and decreased relapse of chronic recurrent viral mucocutaneous infections caused by human papilloma virus and herpes simplex virus by modulating ROS/RNS balance to decrease oxidative/inflammatory damage, increasing antiviral cytokines, and hence decreasing viral load [37]. Vitamin E concentrations significantly decreased due to retroviral infection, and their supplementations at different fold concentrations increased both hepatic and serum levels of vitamin E in both uninflected and infected mice [38]. Increased serum tocopherol levels lead to decrease in AIDS progression and delay in the onset of AIDS in HIV-1-infected individuals [39]. It has also been seen that retroviral infections induce tissue vitamin E deficiency which could be prevented by melatonin (MLT) and dehydroepiandrosterone (DHEAS) supplementations [40]. Vitamin E plays a crucial role in proper immune response by supplementing increased cell-mediated and humoral immune responses and increased phagocytic activity as it directly interacts with macrophages providing resistance to viral and infectious diseases [41,42].

VITAMIN E INTERVENTION AND IMMUNOMODULATION

Vitamin E is also known to help in enhancing immunity in old age by virtue of its antioxidant potentials [43]. Vitamin E supplementation increases immature T-cell differentiation in thymus which usually decreases with age leading to decreased cellular immunity with age. Hence, vitamin E supplements proper immune function in old age. It also plays an important role in protecting cellular membranes, especially immune cells by inhibiting peroxidation of polyunsaturated fatty acids (PUFA). It has also been seen that vitamin E deficiency leads to decreased IL-2 production, B- & T-cell mitogenesis, NK activity, plaque-forming cells (PFC), antibody titer, macrophage, and polymorphonuclear (PMN) phagocytosis and PMN chemotaxis. During vitamin E deficiency, macrophages and neutrophils had high oxygen consumption and H_2O_2 release [44]. Lower vitamin E serum levels lead to increased wheezing among smokers leading to increased respiratory morbidity but no such correlation was seen in non-smokers [45]. Both vitamin E and selenium act as anti-oxidants hence decreasing the risk of infections by increasing resistance to such infections such as respiratory infections caused during COVID-19 [46,47]. It has also been observed that mixed tocopherols are more efficient than α-tocopherol alone as they will target different ranges of receptors [48]. A decrease in eicosatetraenoic acid in mononuclear cells was seen in chronic hepatitis C patients when given vitamin E and C supplementation with interferon alpha-2b and ribavirin combination therapy [49].

NOVEL ASPECTS OF VITAMIN INTERVENTION IN VIRAL DISEASES

Appropriate nutrient intake accounts for the maintenance of immunological equilibrium, in humans and animals. Vitamins, elements, lipids, proteins, and nucleic acids account for major role playing in regulation of cellular and humoral immune responses, since single or multiple deficits of these food components have been shown to cause immune deregulation. For instance, in the course of protein-calorie malnutrition, the major cause of death is represented by bacterial and/or viral infections. Quite interestingly, recent findings outlined some similarities shared between HIV-1 infection and aging in terms of immunological changes. The model of HIV infection may be useful as a futuristic approach for the interpretation of aging mechanisms and possible as a therapeutic intervention.

Starting in early December 2019, the novel coronavirus disease (COVID-19) has caused a global pandemic. Many aspects of its pathogenesis, infection, and related clinical consequences are still unclear. Early diagnosis and dynamic monitoring of prognostic factors are important to improve the ability to manage COVID-19 infection. A study was aimed to provide an account of the role played by vitamins C and D on the onset, progression, and severity of COVID-19. In March 2022, the main online databases were accessed. All the articles that investigate the possible role of vitamins C and D on COVID-19 susceptibility, severity, and progression were considered. The current evidence on vitamin C and D supplementation in patients with COVID-19 infection is inconsistent and still unclear. In some studies, vitamins were used as co-adjuvant of a formal experimental therapy, while in others as main treatment. The major concern is that some of the recent clinical innovations in vitamin interventional approaches to combat infectious diseases was ethnicity and hospital setting (inpatient/outpatient). Moreover, there was no consensus between studies in administration protocol; high heterogeneity in dosage, administration, and duration of the treatment were evident. Finally, some studies administered vitamins before and/or during COVID-19 infection, in patients with different risk factors and infection severity. While waiting to develop a targeted, safe, and effective therapy, it is important to investigate individual predisposition and proper disease management. However, there is a lack of evidence-based guidelines which recommend vitamin C and D supplementation in patients with COVID-19, and results from high-quality randomized controlled trials (RCTs) are inconsistent. Current investigations so far are mostly observational and include a relatively small sample size which can lead to biased results. Large-scale multicenter studies are therefore needed. The current pandemic forced us to introspect and revisit our armamentarium of medicinal agents which could be life-saving in emergency situations. Oxygen diffusion-enhancing compounds represent one such class of potential therapeutic agents, particularly in ischemic conditions. As rewarding as the name suggests, these agents, represented by the most advanced and first-in-class molecule, trans-sodium crocetinate (TSC), are the subject of intense clinical investigation, including Phase 1b/2b clinical trials for COVID-19. Being a successor of a natural product, crocetin, TSC is being investigated for various cancers as a radiosensitizer owing to its oxygen diffusion enhancement capability. The unique properties of TSC make it a promising therapeutic agent for various ailments such as hemorrhagic shock, stroke, and heart attack, among others. The recent literature focusing on the delivery aspects of these compounds is covered as well to paint the complete picture to the curious reader. Given the potential TSC holds as a first-in-class agent, small- and/or macromolecular therapeutics based on the core concept of improved oxygen diffusion from blood to the surrounding tissues, where it is needed the most, will be developed in future and satisfy the unmet medical need for many diseases and disorders [50–52].

CLINICAL TRIALS FOR VITAMIN THERAPIES AGAINST VIRAL INFECTIONS

A randomized clinical trial was held at a tertiary Vanderbilt University hospital from October 2020 to November 2021, and 255 healthcare workers (age 47 ± 12 years, 199 women) were given at least two months of vitamin D3 supplementation with a control group of 2,827 workers. Vitamin D3 5000 IU supplementation was associated with a lower risk of influenza-like illness (ILI) and a lower incidence rate for non-COVID-19 ILI, although COVID-19 ILI incidence was not statistically different. Concluding that daily supplementation of 5,000 IU vitamin D3 reduces ILI in healthcare workers [53], another trial investigates the effect of a 5,000 IU versus 1,000 IU daily oral vitamin D3 supplementation on patients with COVID-19 and their recovery among 69 COVID-19-positive adults. A significant increase in serum 25(OH)D levels, decrease in Body Mass Index (BMI) and IL-6 levels, and shorter recovery time were seen in the 5,000 IU group as compared to 1,000 IU. Also, recommending the use of 5,000 IU vitamin D3 as an adjuvant therapy for COVID-19 patients [54], a combinatorial therapy for COVID-19 patients went under clinical trials to test the effects of calcifediol along with best available therapy of hydroxychloroquine and azithromycin showing a reduced need for ICU treatment [55]. Also, a clinical trial testing the effect of intravenously administered vitamin C on COVID-19 hospitalized patients and their improvement for organ support-free days yielded probability for efficacy of 2.9% among 1,022 patients who were not critically ill and 8.6% among 1,568 critically ill patients related to the odds of improvement for organ support-free days [56]. It has also been seen through clinical trials with over 94 candidates showing that L-arginine plus vitamin C supplementation improved conditions in adults with long COVID-19 such as reduced fatigue, improved muscle strength, and walking performance, and their synergistic effects helped in recovery of COVID-19 patients. L-arginine supplementation has been seen helpful in aerobic and anaerobic performance in athletes, decrease in pulmonary arterial pressure and vascular resistance, and increase in oxygen consumption [57]. No beneficial effect of vitamin A on acute respiratory syncytial virus (RSV) infection among 239 children was seen in a randomized placebo-controlled trial. Also, showing an adverse effect on a group of children leads to longer stay at hospital [58]. In a clinical trial of 94 people, a positive effect of combination of 1,000 mg vitamin C plus 10 mg zinc on common cold was seen. A significant reduction in rhinorrhea was seen as compared to placebo on 5 days of treatment with a quicker recovery and tolerance to the treatment [59]. A clinical trial conducted on 1,300 healthy children in Vietnam to see the effect of vitamin D on influenza infection showed non-significant effect on incidence of influenza but rather a significant reduction in non-influenza respiratory viral infections [60].

Epidemiological studies of 7,807 individuals on their vitamin D status showed increased risk of COVID-19 infection and hospitalization with low vitamin D status. Hence, more clinical trials of vitamin D supplementation are recommended [61]. Over 39 randomized controlled trials with 16,797 candidates were evaluated and concluded that vitamin D supplementation plays a major role in reduction of viral respiratory tract infections (RTIs) particularly in those with vitamin D deficiency [62].

REFERENCES

1. Stephensen, C. B., & Lietz, G. (2021). Vitamin A in resistance to and recovery from infection: relevance to SARS-CoV2. *The British Journal of Nutrition*, *126*(11), 1663–1672. doi:10.1017/S0007114521000246

2. Hemilä, H. (2017). Vitamin C and infections. *Nutrients*, *9*(4), 339. doi:10.3390/nu9040339

3. Batista, K. S., Cintra, V. M., Lucena, P. A. F., Manhães-de-Castro, R., Toscano, A. E., Costa, L. P.,..., & Aquino, J. de S. (2022). The role of vitamin B12 in viral infections: a comprehensive review of its relationship with the muscle-gut-brain axis and implications for SARS-CoV-2 infection. *Nutrition Reviews*, *80*(3), 561–578. doi:10.1093/nutrit/nuab092

4. Bae, M., & Kim, H. (2020). Mini-review on the roles of vitamin C, vitamin D, and selenium in the immune system against COVID-19. *Molecules*, *25*(22), 5346. doi:10.3390/molecules25225346

5. Brown, T. T., & McComsey, G. A. (2010). Association between initiation of antiretroviral therapy with efavirenz and decreases in 25-hydroxyvitamin D. *Antiviral Therapy*, *15*(3), 425–429. doi:10.3851/IMP1502

6. Sudfeld, C. R., Manji, K. P., Duggan, C. P., Aboud, S., Muhihi, A., Sando, D. M.,..., & Fawzi, W. W. (2017). Effect of maternal vitamin D3 supplementation on maternal health, birth outcomes, and infant growth among HIV-infected Tanzanian pregnant women: study protocol for a randomized controlled trial. *Trials*, *18*(1), 411. doi:10.1186/s13063-017-2157-3

7. Yin, M. (2012). Vitamin D, bone, and HIV infection. *Topics in Antiviral Medicine*, *20*(5), 168–172. Retrieved from https://www.ncbi.nlm.nih.gov/pubmed/23363695

8. Fiorino, S., Conti, F., Gramenzi, A., Loggi, E., Cursaro, C., Di Donato, R.,..., & Andreone, P. (2011). Vitamins in the treatment of chronic viral hepatitis. *The British Journal of Nutrition*, *105*(7), 982–989. doi:10.1017/S0007114510004629

9. Xu, Y., Bai, L., Liu, Y., Liu, Y., Xu, T., Xie, S.,..., & Xu, D. (2010). A new triterpenoid saponin from Pulsatilla cernua. *Molecules*, *15*(3), 1891–1897. doi:10.3390/molecules15031891

10. Schleicher, R. L., Carroll, M. D., Ford, E. S., & Lacher, D. A. (2009). Serum vitamin C and the prevalence of vitamin C deficiency in the United States: 2003-2004 National Health and Nutrition Examination Survey (NHANES). *The American Journal of Clinical Nutrition*, *90*(5), 1252–1263. doi:10.3945/ajcn.2008.27016

11. Fiorino, S., Bacchi-Reggiani, L., Sabbatani, S., Grizzi, F., di Tommaso, L., Masetti, M.,..., & Pession, A. (2014). Possible role of tocopherols in the modulation of host microRNA with potential antiviral activity in patients with hepatitis B virus-related persistent infection: a systematic review. *The British Journal of Nutrition*, *112*(11), 1751–1768. doi:10.1017/S0007114514002839

12. Kim, Y., Kim, H., Bae, S., Choi, J., Lim, S. Y., Lee, N.,..., & Lee, W. J. (2013). Vitamin C is an essential factor on the anti-viral immune responses through the production of interferon-α/β at the initial stage of influenza A virus (H3N2) infection. *Immune Network*, *13*(2), 70–74. doi:10.4110/in.2013.13.2.70

13. Mora, J. R., Iwata, M., & von Andrian, U. H. (2008). Vitamin effects on the immune system: vitamins A and D take centre stage. *Nature Reviews. Immunology*, *8*(9), 685–698. doi:10.1038/nri2378

14. Cantorna, M. T., & Mahon, B. D. (2005). D-hormone and the immune system. *The Journal of Rheumatology. Supplement*, *76*, 11–20. Retrieved from https://www.ncbi.nlm.nih.gov/pubmed/16142846

15. Abu-Amer, Y., & Bar-Shavit, Z. (1994). Regulation of TNF-alpha release from bone marrow-derived macrophages by vitamin D. *Journal of Cellular Biochemistry*, *55*(4), 435–444. doi:10.1002/jcb.240550404

16. Staeva-Vieira, T. P., & Freedman, L. P. (2002). 1,25–dihydroxyvitamin D3 inhibits IFN-gamma and IL-4 levels during in vitro polarization of primary murine CD4+ T cells. *The Journal of Immunology, 168*(3), 1181–1189. doi:10.4049/jimmunol.168.3.1181

17. Penna, G., & Adorini, L. (2000). 1 Alpha,25-dihydroxyvitamin D3 inhibits differentiation, maturation, activation, and survival of dendritic cells leading to impaired alloreactive T cell activation. *The Journal of Immunology, 164*(5), 2405–2411. doi:10.4049/jimmunol.164.5.2405

18. Farquharson, C., Law, A. S., Seawright, E., Burt, D. W., & Whitehead, C. C. (1996). The expression of transforming growth factor-beta by cultured chick growth plate chondrocytes: differential regulation by 1,25-dihydroxyvitamin D3. *The Journal of Endocrinology, 149*(2), 277–285. doi:10.1677/joe.0.1490277

19. de la Torre, M. S., Torres, C., Nieto, G., Vergara, S., Carrero, A. J., Macías, J.,..., & Fibla, J. (2008). Vitamin D receptor gene haplotypes and susceptibility to HIV-1 infection in injection drug users. *The Journal of Infectious Diseases, 197*(3), 405–410. doi:10.1086/525043

20. Motsinger-Reif, A. A., Antas, P. R. Z., Oki, N. O., Levy, S., Holland, S. M., & Sterling, T. R. (2010). Polymorphisms in IL-1beta, vitamin D receptor Fok1, and Toll-like receptor 2 are associated with extrapulmonary tuberculosis. *BMC Medical Genetics, 11*(1), 37. doi:10.1186/1471-2350-11-37

21. Bellamy, R., Ruwende, C., Corrah, T., McAdam, K. P., Thursz, M., Whittle, H. C., & Hill, A. V. (1999). Tuberculosis and chronic hepatitis B virus infection in Africans and variation in the vitamin D receptor gene. *The Journal of Infectious Diseases, 179*(3), 721–724. doi:10.1086/314614

22. Saito, M., Eiraku, N., Usuku, K., Nobuhara, Y., Matsumoto, W., Kodama, D.,..., & Osame, M. (2005). ApaI polymorphism of vitamin D receptor gene is associated with susceptibility to HTLV-1-associated myelopathy/tropical spastic paraparesis in HTLV-1 infected individuals. *Journal of the Neurological Sciences, 232*(1–2), 29–35. doi:10.1016/j.jns.2005.01.005

23. Ovsyannikova, I. G., Haralambieva, I. H., Dhiman, N., O'Byrne, M. M., Pankratz, V. S., Jacobson, R. M., & Poland, G. A. (2010). Polymorphisms in the vitamin A receptor and innate immunity genes influence the antibody response to rubella vaccination. *The Journal of Infectious Diseases, 201*(2), 207–213. doi:10.1086/649588

24. Ovsyannikova, I. G., Dhiman, N., Haralambieva, I. H., Vierkant, R. A., O'Byrne, M. M., Jacobson, R. M., & Poland, G. A. (2010). Rubella vaccine-induced cellular immunity: evidence of associations with polymorphisms in the Toll-like, vitamin A and D receptors, and innate immune response genes. *Human Genetics, 127*(2), 207–221. doi:10.1007/s00439-009-0763-1

25. Ovsyannikova, I. G., Haralambieva, I. H., Vierkant, R. A., O'Byrne, M. M., Jacobson, R. M., & Poland, G. A. (2012). Effects of vitamin A and D receptor gene polymorphisms/haplotypes on immune responses to measles vaccine. *Pharmacogenetics and Genomics, 22*(1), 20–31. doi:10.1097/FPC.0b013e32834df186

26. Mahon, B. D., Gordon, S. A., Cruz, J., Cosman, F., & Cantorna, M. T. (2003). Cytokine profile in patients with multiple sclerosis following vitamin D supplementation. *Journal of Neuroimmunology, 134*(1–2), 128–132. doi:10.1016/s0165-5728(02)00396-x

27. Peelen, E., Knippenberg, S., Muris, A.-H., Thewissen, M., Smolders, J., Tervaert, J. W. C.,..., & Damoiseaux, J. (2011). Effects of vitamin D on the peripheral adaptive immune system: a review. *Autoimmunity Reviews, 10*(12), 733–743. doi:10.1016/j.autrev.2011.05.002

28. Smolders, J., Menheere, P., Kessels, A., Damoiseaux, J., & Hupperts, R. (2008). Association of vitamin D metabolite levels with relapse rate and disability in multiple sclerosis. *Multiple Sclerosis, 14*(9), 1220–1224. doi:10.1177/1352458508094399

29. Mowry, E. M., Krupp, L. B., Milazzo, M., Chabas, D., Strober, J. B., Belman, A. L.,..., & Waubant, E. (2010). Vitamin D status is associated with relapse rate in pediatric-onset multiple sclerosis. *Annals of Neurology, 67*(5), 618–624. doi:10.1002/ana.21972

30. Simpson, S., Jr, Taylor, B., Blizzard, L., Ponsonby, A.-L., Pittas, F., Tremlett, H.,..., & van der Mei, I. (2010). Higher 25-hydroxyvitamin D is associated with lower relapse risk in multiple sclerosis. *Annals of Neurology, 68*(2), 193–203. doi:10.1002/ana.22043

31. Ascherio, A., Munger, K. L., & Lünemann, J. D. (2012). The initiation and prevention of multiple sclerosis. *Nature Reviews. Neurology, 8*(11), 602–612. doi:10.1038/nrneurol.2012.198

32. Teymoori-Rad, M., Mozhgani, S.-H., Zarei-Ghobadi, M., Sahraian, M. A., Nejati, A., Amiri, M. M.,..., & Marashi, S. M. (2019). Integrational analysis of miRNAs data sets as a plausible missing linker between Epstein-Barr virus and vitamin D in relapsing remitting MS patients. *Gene, 689*, 1–10. doi:10.1016/j.gene.2018.12.004

33. Wang, Y., Huang, D. S., Wood, S., & Watson, R. R. (1995). Modulation of immune function and cytokine production by various levels of vitamin E supplementation during murine AIDS. *Immunopharmacology, 29*(3), 225–233. doi:10.1016/0162-3109(95)00061-w

34. Han, S. N., Wu, D., Ha, W. K., Beharka, A., Smith, D. E., Bender, B. S., & Meydani, S. N. (2000). Vitamin E supplementation increases T helper 1 cytokine production in old mice infected with influenza virus. *Immunology, 100*(4), 487–493. doi:10.1046/j.1365-2567.2000.00070.x

35. Tang, A. M., Graham, N. M., Semba, R. D., & Saah, A. J. (1997). Association between serum vitamin A and E levels and HIV-1 disease progression. *AIDS, 11*(5), 613–620. doi:10.1097/00002030-199705000-00009

36. Mahmood, S., Yamada, G., Niiyama, G., Kawanaka, M., Togawa, K., Sho, M.,..., & Yodoi, J. (2003). Effect of vitamin E on serum aminotransferase and thioredoxin levels in patients with viral hepatitis C. *Free Radical Research, 37*(7), 781–785. doi:10.1080/1071576031000102141

37. De Luca, C., Kharaeva, Z., Raskovic, D., Pastore, P., Luci, A., & Korkina, L. (2012). Coenzyme Q(10), vitamin E, selenium, and methionine in the treatment of chronic recurrent viral mucocutaneous infections. *Nutrition, 28*(5), 509–514. doi:10.1016/j.nut.2011.08.003

38. Wang, Y., Huang, D. S., Wood, S., & Watson, R. R. (1995). Modulation of immune function and cytokine production by various levels of vitamin E supplementation during murine AIDS. *Immunopharmacology, 29*(3), 225–233. doi:10.1016/0162-3109(95)00061-w

39. Tang, A. M., Graham, N. M., Semba, R. D., & Saah, A. J. (1997). Association between serum vitamin A and E levels and HIV-1 disease progression. *AIDS, 11*(5), 613–620. doi:10.1097/00002030-199705000-00009

40. Zhang, Z., Araghi-Niknam, M., Liang, B., Inserra, P., Ardestani, S. K., Jiang, S.,..., & Watson, R. R. (1999). Prevention of immune dysfunction and vitamin E loss by dehydroepiandrosterone and melatonin supplementation during murine retrovirus infection. *Immunology, 96*(2), 291–297. doi:10.1046/j.1365-2567.1999.00628.x

41. Seidman, E. (2012). An emerging action science of social settings. *American Journal of Community Psychology, 50*(1–2), 1–16. doi:10.1007/s10464-011-9469-3

42. Odeleye, O. E., & Watson, R. R. (1991). The potential role of vitamin E in the treatment of immunologic abnormalities during acquired immune deficiency syndrome. *Progress in Food & Nutrition Science, 15*(1–2), 1–19. Retrieved from https://www.ncbi.nlm.nih.gov/pubmed/1887063

43. Tafe, L. J., Janjigian, Y. Y., Zaidinski, M., Hedvat, C. V., Hameed, M. R., Tang, L. H.,..., & Barbashina, V. (2011). Human epidermal growth factor receptor 2 testing in gastroesophageal cancer: correlation between immunohistochemistry and fluorescence in situ hybridization. *Archives of Pathology & Laboratory Medicine, 135*(11), 1460–1465. doi:10.5858/arpa.2010-0541-OA

44. Moriguchi, S., & Muraga, M. (2000). Vitamin E and immunity. *Vitamins and Hormones, 59*, 305–336. Retrieved from https://www.ncbi.nlm.nih.gov/pubmed/10714244

45. Salo, P. M., Mendy, A., Wilkerson, J., Molsberry, S. A., Feinstein, L., London, S. J.,..., & Zeldin, D. C. (2022). Serum antioxidant vitamins and respiratory morbidity and mortality: a pooled analysis. *Respiratory Research, 23*(1), 150. doi:10.1186/s12931-022-02059-w

46. Kieliszek, M., & Lipinski, B. (2020). Selenium supplementation in the prevention of coronavirus infections (COVID-19). *Medical Hypotheses, 143*(109878), 109878. doi:10.1016/j.mehy.2020.109878

47. Shakoor, H., Feehan, J., Al Dhaheri, A. S., Ali, H. I., Platat, C., Ismail, L. C.,..., & Stojanovska, L. (2021). Immune-boosting role of vitamins D, C, E, zinc, selenium and omega-3 fatty acids: could they help against COVID-19? *Maturitas, 143*, 1–9. doi:10.1016/j.maturitas.2020.08.003

48. Liu, M., Wallin, R., Wallmon, A., & Saldeen, T. (2002). Mixed tocopherols have a stronger inhibitory effect on lipid peroxidation than alpha-tocopherol alone. *Journal of Cardiovascular Pharmacology, 39*(5), 714–721. Retrieved from https://www.ncbi.nlm.nih.gov/pubmed/11973415

49. Murakami, Y., Nagai, A., Kawakami, T., Hino, K., Kitase, A., Hara, Y.-I.,..., & Okita, M. (2006). Vitamin E and C supplementation prevents decrease of eicosapentaenoic acid in mononuclear cells in chronic hepatitis C patients during combination therapy of interferon alpha-2b and ribavirin. *Nutrition, 22*(2), 114–122. doi:10.1016/j.nut.2005.05.014

50. Amati, L., Cirimele, D., Pugliese, V., Covelli, V., Resta, F., & Jirillo, E. (2003). Nutrition and immunity: laboratory and clinical aspects. *Current Pharmaceutical Design, 9*(24), 1924–1931. doi:10.2174/1381612033454252

51. Migliorini, F., Vaishya, R., Eschweiler, J., Oliva, F., Hildebrand, F., & Maffulli, N. (2022). Vitamins C and D and COVID-19 susceptibility, severity and progression: an evidence based systematic review. *Medicina, 58*(7), 941. doi:10.3390/medicina58070941

52. Shah, H. M., Jain, A. S., Joshi, S. V., & Kharkar, P. S. (2021). Crocetin and related oxygen diffusion-enhancing compounds: review of chemical synthesis, pharmacology, clinical development, and novel therapeutic applications. *Drug Development Research, 82*(7), 883–895. doi:10.1002/ddr.21814

53. van Helmond, N., Brobyn, T. L., LaRiccia, P. J., Cafaro, T., Hunter, K., Roy, S.,..., & Chung, M. K. (2022). Vitamin D3 supplementation at 5000 IU daily for the prevention of influenza-like illness in healthcare workers: a pragmatic randomized clinical trial. *Nutrients, 15*(1), 180. doi:10.3390/nu15010180

54. Sabico, S., Enani, M. A., Sheshah, E., Aljohani, N. J., Aldisi, D. A., Alotaibi, N. H.,..., & Al-Daghri, N. M. (2021). Effects of a 2-week 5000 IU versus 1000 IU vitamin D3 supplementation on recovery of symptoms in patients with mild to moderate covid-19: a randomized clinical trial. *Nutrients, 13*(7), 2170. doi:10.3390/nu13072170

55. Entrenas Castillo, M., Entrenas Costa, L. M., Vaquero Barrios, J. M., Alcalá Díaz, J. F., López Miranda, J., Bouillon, R., & Quesada Gomez, J. M. (2020). Effect of calcifediol treatment and best available therapy versus best available therapy on intensive care unit admission and mortality among patients hospitalized for COVID-19: a pilot randomized clinical study. *The Journal of Steroid Biochemistry and Molecular Biology, 203*(105751), 105751. doi:10.1016/j.jsbmb.2020.105751

56. LOVIT-COVID Investigators, on behalf of the Canadian Critical Care Trials Group, and the REMAP-CAP Investigators, Adhikari, N. K. J., Hashmi, M., Tirupakuzhi Vijayaraghavan, B. K., Haniffa, R., Beane, A.,..., & Lamontagne, F. (2023). Review of Intravenous vitamin C for patients hospitalized with COVID-19: two harmonized randomized clinical trials. *JAMA: The Journal of the American Medical Association, 330*(18), 1745–1759. doi:10.1001/jama.2023.21407

57. Tosato, M., Calvani, R., Picca, A., Ciciarello, F., Galluzzo, V., Coelho-Júnior, H. J.,..., & Gemelli against COVID-19 Post-Acute Care Team. (2022). Effects of l-arginine plus vitamin C supplementation on physical performance, endothelial function, and persistent fatigue in adults with long COVID: a single-blind randomized controlled trial. *Nutrients, 14*(23), 4984. doi:10.3390/nu14234984

58. Bresee, J. S., Fischer, M., Dowell, S. F., Johnston, B. D., Biggs, V. M., Levine, R. S.,..., & Anderson, L. J. (1996). Vitamin A therapy for children with respiratory syncytial virus infection: a multicenter trial in the United States. *The Pediatric Infectious Disease Journal, 15*(9), 777–782. doi:10.1097/00006454-19960 9000-00008

59. Maggini, S., Beveridge, S., & Suter, M. (2012). A combination of high-dose vitamin C plus zinc for the common cold. *The Journal of International Medical Research, 40*(1), 28–42. doi:10.1177/147323001204000104

60. Loeb, M., Dang, A. D., Thiem, V. D., Thanabalan, V., Wang, B., Nguyen, N. B.,..., & Pullenayegum, E. (2019). Effect of vitamin D supplementation to reduce respiratory infections in children and adolescents in Vietnam: a randomized controlled trial. *Influenza and Other Respiratory Viruses, 13*(2), 176–183. doi:10.1111/irv.12615

61. Camargo, C. A., Jr, & Martineau, A. R. (2020). Vitamin D to prevent COVID-19: recommendations for the design of clinical trials. *The FEBS Journal, 287*(17), 3689–3692. doi:10.1111/febs.15534

62. Shokri-Mashhadi, N., Kazemi, M., Saadat, S., & Moradi, S. (2021). Effects of select dietary supplements on the prevention and treatment of viral respiratory tract infections: a systematic review of randomized controlled trials. *Expert Review of Respiratory Medicine, 15*(6), 805–821. doi:10.1080/17476348.2021.191 8546

Abbreviations

AA	Amino acids
ABO	Antigen A, B for ABO-blood grouping
ACE	Angiotensin-converting enzyme 2
AE	Adverse events
AF	Air filtration
AHR	Aryl hydrocarbon receptor
AIDS	Acquired immunodeficiency syndrome
AKT	Protein kinase B
ALI	Acute lung injury
ALT	Alanine aminotransferase
AMDHD	Amidohydrolase domain-containing protein
AMP	Adenosine monophosphate
AMPK	AMP-activated protein kinase
APC	Antigen-presenting cell
AR	Androgen receptor
ARDS	Acute respiratory distress syndrome
ARI	Acute respiratory infection
BALF	Bronchoalveolar lavage fluid
BEBOV	Bundibugyo Ebola virus
BMI	Body mass index
CAT	Computerized axial tomography
CD	Cluster of differentiation
CDC	Centers for Disease Control and Prevention
CITRIS	Center for Information Technology Research in the Interest of Society
CNS	Central nervous system
COPD	Chronic obstructive pulmonary disease
COV	Coronavirus
COVID	Coronavirus disease
CP	Convalescent plasma
CRP	C-reactive protein
CT	Computed tomography
CTD	Connective tissue disease
CVD	Cardiovascular disease
CYP	Cytochrome P450
DA	Dopamine
DAD	Diffuse alveolar damage
DALY	Disability-adjusted life year
DC	Dendritic cell
DENV	Dengue virus
DHCC	Dihydroxycholecalciferol
DHCR	Dehydrocholesterol reductase
DHEAS	Dehydroepiandrosterone sulfate
DHR	Dihydrorhodamine

DIC	Disseminated intravascular coagulation
DNA	Deoxyribonucleic acid
DPP	Dipeptidyl peptidase
EBOV	Ebola virus
EBV	Epstein–Barr virus
EGFR	Epidermal growth factor receptor
EO	Emergency operations
ER	Estrogen receptor
ERK	Extracellular signal-regulated kinase
ES	Emergency services
ETU	Ebola treatment unit
EVD	Ebola virus disease
EVICT	Emergency ventilator-induced COVID-19 therapy
FDA	Food and Drug Administration
GBD	Global burden of diseases
GC	Genomic Consortium
GISRS	Global Influenza Surveillance and Response System
GP	Glycoprotein
GSH	Glutathione (in reduced form)
HA	Hemagglutinin
HAT	Hypoxanthine-aminopterin-thymidine
HBV	Hepatitis B virus
HC	Healthcare
HCV	Hepatitis C virus
HIV	Human immunodeficiency virus
HKU	Hong Kong University
HLA	Human leukocyte antigen
HRV	Human rhinovirus
HTLV	Human T-cell lymphotropic virus
IBV	Infectious bronchitis virus
ICMR	Indian Council of Medical Research
ICU	Intensive care unit
IFN	Interferon
ILI	Influenza-like illness
IL	Interleukin
IMC	Incident Management Center
IMVFD	Intermittent mandatory ventilation with FiO_2 delivery
IRF	Interferon regulatory factor
ISG	Interferon-stimulated gene
IU	International unit
IV	Intravenous
JAK	Janus kinase
JAMA	*Journal of the American Medical Association*
LOVIT	Lung organoid vascular inflammation therapy
LRAT	Lecithin:retinol acyltransferase
MA	Mutual aid
MAP	Mitogen-activated protein
MAPK	Mitogen-activated protein kinase
MARS	Molecular adsorbents recirculating system
MDA	Mass drug administration
MERS	Middle East respiratory syndrome

MGP	Monoclonal antibody-guided plasma
MHC	Major histocompatibility complex
MK	Mobile kitchen
MLT	Medical laboratory technician
MO	Medical officer
MS	Molecular surveillance
MV	Mechanical ventilation
MX	Mixed reality
NA	Nucleic acid
NAD	Nicotinamide adenine dinucleotide
NADH	Nicotinamide adenine dinucleotide (reduced)
NADP	Nicotinamide adenine dinucleotide phosphate
NADPH	Nicotinamide adenine dinucleotide phosphate (reduced)
NCT	Nasopharyngeal carriage testing
NET	Neutrophil extracellular trap
NF	Necrotizing fasciitis
NIC	National immunization campaign
NK	Natural killer
NKT	Natural killer T-cells
NL	Nighttime lockdown
NLR	Nucleotide-binding domain, leucine-rich repeat
NPC	Nasopharyngeal culture
NSP	Non-structural protein
OC	Outbreak control
OH	Occupational health
ONSCOVID	Ontario COVID-19 (referring to the province of Ontario in Canada)
OR	Oxygen requirement
ORF	Open reading frame (a region of a virus's genome that can be translated into proteins)
PBMC	Peripheral blood mononuclear cell
PEP	Post-exposure prophylaxis
PFC	Pulmonary function capacity
PKC	Protein kinase C
PL	Phospholipid
PLP	Pyridoxal phosphate (a coenzyme form of vitamin B6)
PMN	Polymorphonuclear leukocyte (a type of white blood cell)
PMP	Personal protective equipment
PRKCB	Protein kinase C beta
PRR	Pattern recognition receptor
PUFA	Polyunsaturated fatty acid
QQ	Quarantine and quarantine measures
RAS	Renin-angiotensin system
RBD	Receptor-binding domain
RBV	Ribavirin (an antiviral medication)
RDA	Recommended dietary allowance
REBOV	Recombinant Ebola virus
RIG	Respiratory immunoglobulin
RNA	Ribonucleic acid
RNS	Reactive nitrogen species
ROS	Reactive oxygen species
RSV	Respiratory syncytial virus
RTC	Replication transcription complex

RV	Rhinovirus (a common cause of the common cold)
RXRA	Retinoid X receptor alpha
SARI	Severe acute respiratory infection
SARS	Severe acute respiratory syndrome
SD	Social distancing
SEBOV	Sudan Ebolavirus
SEC	Spike (S) protein ectodomain cleavage
SNP	Single nucleotide polymorphism
SOFA	Sequential organ failure assessment
STAT	Signal transducer and activator of transcription
SVR	Sustained virologic response
TAFV	Taï forest virus
TB	Tuberculosis
TGF	Transforming growth factor
TLR	Toll-like receptor
TMPRSS	Transmembrane protease serine
TNF	Tumor necrosis factor
TOP	Treatment optimization platform
TRS	Transcription regulatory sequence
TRX	Thioredoxin
TSC	Targeted symptom control
TTP	Thrombotic thrombocytopenic purpura, thymine triphosphate
URI	Upper respiratory infection
USA	United States of America
USSR	Union of Soviet Socialist Republics (Erstwhile – no longer exists)
UV	Ultraviolet (referring to UV light, which can be used for disinfection)
UVA	Ultraviolet type A
UVB	Ultraviolet type B
VDD	Vitamin D deficiency
VDR	Vitamin D receptor
VE	Vitamin E
VICTAS	Vitamin C, thiamine, and steroids in sepsis
VIT	Vaccine-induced thrombotic thrombocytopenia
WHO	World Health Organization (a global health agency)
YLD	Years lived with disability
YLL	Years of life lost
ZEBOV	Zaire Ebola virus (a strain of the Ebola virus)

Glossary

ACE2 (angiotensin-converting enzyme 2)   A cellular receptor utilized by several coronaviruses, including SARS-CoV-2, for viral entry into host cells, particularly abundant in the respiratory and cardiovascular systems.

Acute respiratory distress syndrome (ARDS)   A severe lung condition characterized by inflammation and fluid buildup in the lungs, leading to respiratory failure, commonly observed in severe cases of viral pneumonia, including COVID-19.

Adaptive immunity   The branch of the immune system that develops specific responses to particular pathogens, including viruses, upon initial exposure, providing long-lasting protection against reinfection.

Antibiotic   A medication used to treat bacterial infections and not effective against viral infections.

Antigen   A molecule or substance recognized by the immune system as foreign, typically found on the surface of pathogens such as viruses, triggering an immune response.

Antigen test   A diagnostic test that detects the presence of specific viral antigens in respiratory specimens, providing rapid results for the diagnosis of acute viral infections, such as COVID-19.

Antiviral   Medications or treatments designed to inhibit the replication or activity of viruses, used to prevent or treat viral infections.

Asymptomatic   Having a viral infection or disease without displaying any symptoms or signs of illness.

Attenuated vaccine   A type of vaccine containing live, weakened viruses or bacteria that stimulate an immune response without causing severe illness, used to confer long-lasting immunity against viral diseases with reduced risk of adverse effects.

B cell   A type of white blood cell that produces antibodies and plays a crucial role in the adaptive immune response to viral infections.

Contact tracing   The process of identifying and monitoring individuals who have been in close contact with an infected person to prevent the spread of infectious diseases such as COVID-19.

Coronavirus   A large family of viruses that cause illnesses ranging from the common cold to more severe diseases such as COVID-19, characterized by the presence of crown-like spikes on their surface.

Convalescent plasma   Blood plasma collected from individuals who have recovered from a viral infection and contains antibodies against the virus, used as a passive immunization therapy to treat or prevent severe cases of the same infection in others.

Coronavirus disease 2019 (COVID-19)   An infectious disease caused by the novel coronavirus SARS-CoV-2, first identified in Wuhan, China, in late 2019, characterized by respiratory symptoms ranging from mild to severe, including pneumonia and acute respiratory distress syndrome (ARDS).

Cross-reactivity   The ability of antibodies or immune cells to recognize and respond to similar epitopes or antigens present in different strains or species of viruses, contributing to broad-spectrum immunity against related viral infections.

Cytokine Signaling proteins released by immune cells in response to infection, inflammation, or injury, playing a key role in regulating immune responses to viral infections.

Cytopathic effect Structural or functional changes observed in host cells following viral infection, resulting from viral replication, cellular damage, or immune response, commonly used to assess viral infectivity and pathogenicity in vitro.

Dexamethasone A corticosteroid medication with anti-inflammatory properties, used to reduce inflammation and alleviate symptoms in severe cases of viral infections, including COVID-19-associated cytokine storms and respiratory complications.

Droplet precautions Infection control measures aimed at preventing the transmission of respiratory infections through respiratory droplets, including wearing masks, gloves, and goggles.

Efficacy The ability of a vaccine or treatment to produce the desired effect, such as preventing infection, reducing disease severity, or improving clinical outcomes, often expressed as a percentage in clinical trials.

Endemic The constant presence of a disease or infection within a specific geographic area or population group, such as the seasonal flu in certain regions.

Epidemic A sudden increase in the number of cases of a disease within a specific community or region, exceeding what is normally expected.

Epidemiology The study of the distribution and determinants of diseases in populations, including the incidence, prevalence, risk factors, and patterns of transmission, crucial for understanding and controlling infectious diseases.

Fomite Inanimate objects or surfaces that can harbor and transmit infectious agents, including viruses, contributing to the spread of viral diseases.

Genome The complete set of genetic material (DNA or RNA) present in an organism, including viruses, which encodes all the information necessary for their replication and function.

Herd immunity The indirect protection from infectious diseases that occurs when a large percentage of a population becomes immune to the disease, either through vaccination or prior infection, reducing the spread of the virus within the community.

Herd immunity threshold The proportion of the population that needs to be immune to a specific infectious agent, either through vaccination or prior infection, to prevent sustained transmission and achieve herd immunity.

Host cell A cell infected by a virus, providing the necessary machinery and resources for viral replication and propagation.

Host range The range of species or cells susceptible to infection by a particular virus, determined by factors such as viral tropism, receptor specificity, and host immune response, influencing the epidemiology and zoonotic potential of the virus.

Hydroxychloroquine An antimalarial and immunomodulatory medication investigated for its potential therapeutic effects in viral infections, including COVID-19, despite conflicting evidence regarding its efficacy and safety.

Hygiene hypothesis The theory that reduced exposure to microbes early in life may lead to an increased risk of developing allergies and autoimmune diseases, suggesting a balance between immune stimulation and regulation.

Immunocompromised Having a weakened or impaired immune system, making individuals more susceptible to infections, including viral diseases.

Influenza A contagious viral infection caused by influenza viruses, characterized by respiratory symptoms such as fever, cough, sore throat, and muscle aches, with seasonal epidemics occurring annually.

Interferon Signaling proteins released by host cells in response to viral infection, inducing antiviral defenses to inhibit viral replication and spread.

Immune response The coordinated series of physiological and cellular events initiated by the immune system to recognize, neutralize, and eliminate pathogens, including viruses, and establish immunological memory for future protection.

Incubation period The interval between exposure to a pathogen, such as a virus, and the onset of symptoms or signs of infection, representing the time required for the virus to replicate and establish infection within the host.

Innate immunity The non-specific defense mechanisms that provide immediate protection against infections, including physical barriers, cellular components, and soluble mediators, serving as the first line of defense against viral invaders.

Intensive care unit (ICU) A specialized medical unit equipped to provide comprehensive care for critically ill patients, including those with severe viral infections, requiring advanced monitoring, ventilation, and life support interventions.

Interleukin A group of cytokines produced by immune cells, including T-cells, macrophages, and dendritic cells, involved in regulating inflammatory responses, immune cell activation, and communication during viral infections.

Latent infection A persistent viral infection characterized by periods of viral dormancy or inactivity within host cells, followed by reactivation and recurrent symptoms.

Lymphocyte A type of white blood cell involved in the adaptive immune response, including B-cells, T-cells, and natural killer (NK) cells, crucial for recognizing and eliminating viral infections.

Lockdown A public health measure implemented to restrict movement, social interactions, and non-essential activities within a community or region, aimed at controlling the spread of infectious diseases during outbreaks or pandemics.

Messenger RNA (mRNA) A type of RNA molecule synthesized from DNA templates in the cell nucleus and subsequently translated into protein by ribosomes in the cytoplasm, serving as the genetic blueprint for protein synthesis, including viral proteins.

Mortality rate The proportion of deaths within a population due to a specific disease or condition, often expressed as a percentage.

Mutation A change in the genetic material (DNA or RNA) of a virus, potentially altering its characteristics, including virulence, transmissibility, and resistance to treatments or vaccines.

Mutation rate The frequency at which genetic mutations occur within a viral population over time, influenced by factors such as replication fidelity, selection pressure, and genome stability, shaping viral evolution and diversity.

Monoclonal antibody Laboratory-produced antibodies designed to target specific antigens, including viral proteins, used for diagnostic, therapeutic, or prophylactic purposes against viral infections, such as COVID-19 monoclonal antibody therapies.

Non-pharmaceutical interventions (NPIs) Public health measures aimed at reducing the transmission of infectious diseases, including hand hygiene, physical distancing, mask-wearing, and quarantine.

Nucleic acid test (NAT)   Diagnostic tests used to detect and amplify viral genetic material (RNA or DNA) from patient samples, providing sensitive and specific detection of viral infections, including PCR-based tests for COVID-19.

Nucleoside analog   Synthetic compounds structurally similar to nucleosides, incorporated into viral nucleic acids during replication, interfering with viral polymerase activity and inhibiting viral replication, used as antiviral drugs in the treatment of viral infections.

Outbreak   A localized occurrence of a disease within a specific community, region, or population group, typically larger than sporadic cases but smaller than an epidemic.

Pathogen   An infectious agent, such as a virus, bacterium, fungus, or parasite, capable of causing disease in a host organism.

PCR (polymerase chain reaction)   A laboratory technique used to amplify and detect specific segments of DNA or RNA, widely used for diagnosing viral infections, including COVID-19.

Pneumonia   An inflammatory condition of the lungs often caused by viral infections, characterized by symptoms such as cough, fever, chest pain, and difficulty breathing.

Pharmacokinetics   The study of the absorption, distribution, metabolism, and excretion of drugs or medications within the body, influencing their efficacy, toxicity, and dosing regimens in the treatment of viral infections.

PPE (personal protective equipment)   Specialized clothing or equipment worn by healthcare workers and individuals to protect against exposure to infectious agents, including gloves, masks, goggles, and gowns.

Prevalence   The proportion of individuals in a population who have a particular disease or condition at a specific point in time, often expressed as a percentage.

Prophylaxis   The administration of preventive measures or treatments to reduce the risk of acquiring a specific disease or infection, such as vaccination or antiviral medications.

Quarantine   The isolation of individuals who have been exposed to a contagious disease but are not yet symptomatic, to prevent the spread of the infection to others.

Receptor   A molecule on the surface of a host cell that interacts with viral proteins, facilitating viral attachment, entry, and infection.

Recombination   The process by which genetic material from different viruses mixes and recombines, leading to the generation of novel viral strains with unique characteristics.

Replication   The process by which viruses replicate their genetic material and produce new viral particles within host cells, essential for viral spread and infection.

Reservoir   A natural habitat or host organism where a virus resides and can potentially replicate, serving as a source of infection for other susceptible hosts.

Reverse transcription polymerase chain reaction (RT-PCR)   A laboratory technique used to detect and quantify RNA molecules, including viral RNA, by converting RNA into DNA and amplifying specific DNA sequences.

Severe   A term used to describe the intensity or severity of a disease or infection, often referring to life-threatening complications or conditions.

Transmission   The process by which viruses are spread from one host to another, either directly (e.g., through respiratory droplets) or indirectly (e.g., via contaminated surfaces).

Vaccination The administration of vaccines to stimulate the immune system and provide protection against specific infectious diseases, including viral infections, by inducing immune memory and antibody production.

Variant A subtype or strain of a virus that has distinct genetic characteristics compared to the original virus, potentially affecting transmissibility, virulence, and immune evasion.

Viral load The amount of virus present in an infected person's body, typically measured by the quantity of viral genetic material (RNA or DNA) detected in samples such as blood, saliva, or respiratory secretions.

Vitamins Vitamins are essential organic compounds that are necessary for normal growth, development, and physiological functioning in humans and many other organisms. These micronutrients are required in small quantities, and the body cannot synthesize them in sufficient amounts, necessitating their intake through diet or supplementation.

Vitamins (water-soluble) These vitamins dissolve in water and are not stored in significant amounts in the body. They include vitamin C and the B-complex vitamins (such as B1, B2, B3, B5, B6, B7, B9, and B12). Water-soluble vitamins are typically excreted in urine if consumed in excess, so they need to be replenished regularly through diet.

Vitamins (fat-soluble) These vitamins are soluble in fats and oils and can be stored in the body's fatty tissues and liver. They include vitamins A, D, E, and K. Fat-soluble vitamins are absorbed along with dietary fats and are less readily excreted, so excessive intake can lead to toxicity.

Vitamin C (ascorbic acid) A water-soluble vitamin with antioxidant properties that plays a crucial role in immune function and may help reduce the severity and duration of viral infections.

Vitamin D A fat-soluble vitamin synthesized by the body in response to sunlight exposure, important for maintaining immune function and reducing the risk of respiratory infections.

Zoonotic Referring to diseases that can be transmitted from animals to humans, including Ebola and COVID-19, among many other viral infections.

Index

Note: **Bold** page numbers refer to tables and *italic* page numbers refer to figures.

Printed in the United States
by Baker & Taylor Publisher Services